DOWN AND OUT IN
SILICON VALLEY

the high cost of the
HIGH–TECH DREAM

DOWN AND OUT IN
SILICON VALLEY

MEL KRANTZLER, Ph.D.
PATRICIA BIONDI KRANTZLER, M.A.

Prometheus Books
59 John Glenn Drive
Amherst, New York 14228-2197

Published 2002 by Prometheus Books

Inquiries should be addressed to
Prometheus Books
59 John Glenn Drive
Amherst, New York 14228–2197
VOICE: 716–691–0133, ext. 207
FAX: 716–564–2711
WWW.PROMETHEUSBOOKS.COM

06 05 04 03 02   5 4 3 2 1

Library of Congress Cataloging-in-Publication Data forthcoming

ISBN 1–57392–926–3

Printed in Canada

CONTENTS

ACKNOWLEDGMENT

We greatly value the wholehearted, courageous contribution of the many workers in all phases of Silicon Valley industry who gave their time, effort, and honesty to make this book possible.

The dedication of Dale Hurst, our word processing expert, to the technical shaping of our manuscript was invaluable.

INTRODUCTION

S ilicon Valley is a fifteen-city geographical sprawl located in northern California. If that were its only identity, it would not have the all-encompassing relevance to American life it possesses today. It is, of course, the area which has defined the time we live in as the Information Age—the age of the microchip, the personal computer, the Internet and the World Wide Web. No individual or government or business is immune to its massive impact on human behavior. Without the cutting-edge technological hardware initially created in these relatively small West-Coast cities, the Information Age might still be only a gleam in the eyes of future-oriented scientists, rather than the goliath that strides continents today.

This book is not a technical story of the high-tech industry's origination in Silicon Valley or a tale of rags-to-riches fantasy the media has cultivated until the recent economic debacle. There are no heroes in this book—just fallible human beings victimized by a culture that worships wealth and power to the exclusion of every other aspect of human life.

My wife, Pat, and I have a very special interest in the Silicon Valley phenomenon, since over the past decade we have counseled a wide range of people who are chief executive officers, high-tech experts, technical assistants, or human resource workers. For we live in San Rafael, in Marin County, where Silicon Valley companies have sprawled over an area fifty miles north of Santa Clara County, which gave Silicon Valley its name and birth. We have counseled these people at our center because they have had severe human relationship problems. Somehow, their lives were unraveling in spite of the "good fortune" of their economic situation: major depression, anxiety attacks, drug addictions, the alienation of their children and spouses were why they came to us for personal help at our Creative Divorce, Love & Marriage Counseling Center.

We might add that these people came to us during the height of Silicon Valley prosperity, before the recent economic downturn which has eliminated over a hundred thousand jobs in the area. So even before the current economic recession, the quality of Silicon Valley life was hardly the land of milk-and-honey that the media loved to present to the American public. It is, of course, an area of marvelous photo opportunities: the extraordinary landscapes, the glamorous city of San Francisco, the elegance of Stanford University (the "father" of the high-tech industry), the seductive year-round climate and casual life style.

However, as the great painter René Magritte once remarked, "Everything we see hides another thing." And hidden underneath the fairy-tale media picture of Silicon Valley lies this reality even before the recent torrent of unemployment: It has one of the highest divorce rates in the world; more children who are psychologically disturbed than in less affluent areas; no affordable decent housing, even for those earning fifty thousand dollars a year; and widespread alcohol and drug use.

Aside from the simple geographical designation, Silicon Valley is the name for a psychological obsession found anyplace where people believe that instant fame and fortune can be gained through silicon chips and Web sites, or lotteries or stock-market trading. This dream nourishes itself on an addiction to money, power, and instant gratification. And like heroin and cocaine, it is highly illusory, promising total happiness, but often ending in disarray and despair.

This book is about the quality of American life, about the way we live now, blinded by the brightness of the Silicon Valley dream, for it is that brightness which blinds our eyes to our own potential for becoming truly balanced human beings. The Silicon Valley dream is the American Dream gone ballistic, a place where money and power are the only things that matter in life. It is the market model of the Dow Jones Industrial Average and NASDAQ daily stock quotations that is obsessively reported on all TV news programs and is headlined as a front-page item the moment the market shivers.

This book, therefore, focuses on Silicon Valley as a *universal* psychological phenomenon that we label "The Silicon Valley Syndrome." A syndrome is a group of signs and symptoms that collectively indicate an abnormal condition. We submit that what is considered "normal" in Silicon Valley culture-and-society-at-large is indeed "abnormal" since it stunts our development and our ability to become fully human, beyond the fear and greed that drives that culture. The Silicon Valley Syndrome can be found in Seattle, where

Bill Gates' Microsoft software development predominates; in New York (called "Silicon Alley"); in Austin, Texas, and dozens of other clones of the original valley throughout the world.

However, the Silicon Valley Syndrome has exhilarated itself in its most virulent form in the northern California cities where high-tech originated. It is for this reason our book distills the personal problems Silicon Valley men and women are experiencing, which indicate the ways in which the Silicon Valley dream has become a nightmare for them. It also consists of our decade-long observations of Silicon Valley culture and economic conditions and their psychological impact on the individuals driven by their dreams of money and power. Our book deals with the Silicon Valley experience during the so-called good times of the past decade and the subsequent dot-com debacle we are now witnessing. I have been an economist as well as a psychologist, and Pat has been a social worker as well as a psychologist. The combination of our backgrounds provides the insights for this book.

We have no desire to be downbeat or negative. To the contrary, ours is a cautionary tale that will answer the paramount question that psychologist Howard Gardner asks:

"However appropriate it may be for the economy, the 'market model' is a grossly inadequate model for the rest of the human society. With the decline of religious conviction and the slow pace of changes in the legal code, how can we nurture persons and institutions that can resist a purely market orientation in all spheres of living?"

We wish to thank all of our clients and others we have interviewed for their participation in the preparation of this book. Their privacy is protected since we use fictitious names when they wanted us to do so. The heartfelt honesty of their interviews has been a major gift, enabling us to write this book in its present form.

SILICON VALLEY
THEN AND NOW

I t wasn't always called Silicon Valley. That was the label attached to the Santa Clara Valley in the early 1970s by a reporter who noted the proliferation of microchip companies (the chips were silicon-based) that was beginning to occur during the decade.

However, when we flash back to pre–World War II times, the valley indeed was a valley, not a chaos of antiseptic-looking companies that turned the area into a factory complex. Santa Clara old-timers can recall when it really was a valley in the agricultural sense of the word. There were lush orchards producing cherries, apricots, and a host of other fruits and vegetables. It was an area that proudly proclaimed it was the prune capital of the world. The pace of the cities in the valley matched the peacefulness of the orchards. The air was clear, the water pure, the cars were few, and the noises heard were the pleasant sounds of nature.

This bucolic world began to evanesce with increasing

rapidity after World War II. The seeds of change were planted by Stanford University, located in Palo Alto, the major city along with San Jose and Santa Clara in this region of northern California.

It was Frederick E. Terman, Stanford University's dean of engineering in the postwar period, who laid the groundwork for the subsequent explosive growth of the high-tech companies that dominate the valley today. It was Terman's determination to make Stanford one of the outstanding high-technology institutions in the world, and he succeeded brilliantly in doing so. He was also an encourager of start-up companies by his students. One such became Hewlett-Packard, a leading computer hardware company which began as a small electronics firm in 1939 with an investment of $538.

We have frequently heard Silicon Valley spokespeople rail against government interference and proclaim themselves libertarians, who developed their companies without any outside help whatsoever, yet the reality is exactly the opposite. If Dean Terman were alive today, he would smile at the notion that successful Silicon Valley companies are the result of unfettered private enterprise. For it was Terman who actively solicited military contracts that were the bases for the academic research in microwave electronics, so that by 1949 Stanford University became one of the top three recipients of government research contracts. It was government contracts such as these that enabled the first mass-produced commercial computers to be built in 1951. The first computer was called UNIVAC I, and its first purchaser was the University of California Census Bureau. The cost of this computer was one million dollars. J. Presper

Eckert and John W. Mauchly created UNIVAC I, and during World War II they were also the principal designers of the ENIAC computer specifically built for U.S. Army use. The origin of Silicon Valley industrial development was because of the government's need for sophisticated equipment. It in turn created prosperity during the Korean War through Defense Department orders. Later, this same bastion of support caused a recession in 1990–91 when government orders declined as a result of the end of the Cold War, since the valley's technology, so essential for the national defense, was no longer needed on such a massive scale.

The Hewlett-Packard Company became the first national poster company of Silicon Valley. It is enshrined in the entrepreneurial bible as the glittering example of how you too can become a billionaire by founding a company in a garage with a $538 investment. This garage in Palo Alto has been called "the most familiar icon of humble origins since Lincoln's log cabin." Of course, you might have to be graduate students at one of the most prestigious universities in the world. You would have grown up in a house with comfortably fixed, intellectual parents, like Hewlett and Packard did, and also be outstanding engineering students of a professor like Frederick Terman, who encouraged their start-up and actively supported its need for funding. There was also the right timing to be considered: the Hewlett-Packard start-up began with the United States' entry into World War II, and government contracts fueled its early success. (We will have more to say about government assistance to Silicon Valley enterprises in chapter 3.)

Without the invention of the transistor, associated with

William Shockley as its major creator in 1956, Silicon Valley would probably be called Silicon Gulch. Indeed, the transistor is the heart of modern world industry. Without its existence, the semiconductor would still be a figment of the imagination instead of being the magnificent invention that powers our world today. More than half of all semiconductor sales in the 1960s were made to the U.S. Defense Department, but by 1972 the military accounted for only 12 percent of semiconductor sales, as business applications expanded private sector demand. Start-ups were being funded by venture capitalists starting in the mid-seventies, replacing the military as the primary source of new funding.

There indeed was a major slump in Silicon Valley business between 1986 and 1992. In fact, the valley was then called the "Valley of Deaths" by the many men and women who lost their jobs at that time. However, by the mid-1990s, Silicon Valley began to explode with new start-ups (including the Internet search engine Yahoo, founded in Santa Clara in 1994). It was the period between the mid-1990s and the beginning of the year 2000 that catapulted Silicon Valley into worldwide fame and intense envy. For it was during this short period of time that the Valley became the place where it seemed that the greatest and quickest creation of fortunes in history was taking place. Literally thousands of new companies sprung up overnight, and almost every one of them was presented in the media as the birthplace of numerous millionaires. Stories abounded about how a thousand dollar investment could make you an instant millionaire and that any ordinary person could and should do just that. As we shall see, the euphoria of these five years was transformed into the bleakest despair with the stock-market crash in March 2000—a despair that persists today.

However, the vision of instant, unearned wealth materializing out of a garage start-up by two enterprising graduate students had some basis in reality during those exuberant five years. Just like the gold rush of 1849, when tens of thousands of people migrated west to pick up the gold they believed was just lying in the streets, there was a grain of truth in Silicon Valley's being a land where fabulous fortunes could be made. For there was gold in California in 1849, and a tiny minority of prospectors did become wealthy, but the rest perished economically or physically or both. The same thing happened in the year 2000 in this prosperous region of concentrated technology: a very small minority of start-up millionaires remains after the detritus of thousands of other start-ups have been taken to the garbage dump.

Indeed, there was real opportunity in Silicon Valley that began to become apparent in the mid-1990s. For the recession at the beginning of the decade turned into real prosperity when the Internet and the World Wide Web ushered in the Information Age and the information superhighway, which created an enormous market for personal computers. It was the time when millions of personal computers for home, business, and school became the high-tech purchasing item of choice for the majority of American families, businesses, and schools, and Silicon Valley was the supplier of most of those computers. Hewlett-Packard in Palo Alto and Apple Computers in Cupertino became personal computer names known throughout the world.

In addition, the Valley has a proliferation of high-tech support services and products that enable the area to remain in the forefront of the high-tech business: skilled engineers who come from the major universities in the country (for example, Stanford, Harvard, Yale, and MIT), venture capitalists, consultants, lawyers, advertising agencies, and sales organizations. And Stanford University sits as the towering research center at its heart.

It is the wide variety of high-tech skills working in combination that has made the Valley predominant throughout the world. Other countries have never been able to successfully match the resources Silicon Valley can call upon for making cutting-edge products. Here are some of the major companies in Silicon Valley that use the bellwhether layers in the world industry of high technology:

▼ In the field of finished computers, Hewlett-Packard in Palo Alto, and Apple in Cupertino predominate.

▼ In the field of semiconductors and chip technology (the heart of computers), there are Intel in Santa Clara, AMD in Sunnyvale, and Cypress Semiconductor in San Jose.

▼ In the field of networking, fiber optics, and information storage, there are Cisco Systems (the world giant in the business networking industry) located in San Jose, Juniper Networks in Mountain View, and 3Com in Santa Clara.

▼ In the field of software, there are Oracle and Electronic Arts in Redwood City, Netscape and Intuit in Mountain View, Adobe Systems in San Jose, Broadbase Software in Menlo Park, Pacific Bell in San Francisco, and Autodesk and Lucas Arts in San Rafael.

▼ In the field of the Internet, which allows the general public access to the enormous variety of Internet services, there are Google and Excite @ Home in Redwood City, C-Net in San Francisco, eBay and WebTV and Two in San Jose, C-Net and NB @ Internet and Look Smart in San Francisco, E-trade in Menlo Park, Altavista in Palo Alto, and Yahoo in Santa Clara.

▼ Media organizations with Web sites are KRON-TV, KPIX-TV, SFGate, Red Herring, Tech TV, and Wired Digital in San Francisco; Sunset Publications in Menlo Park; and Knight-Ridder (Mercury News) in San Jose.

There are many, many more companies in the area, such as banking, venture capital and legal organizations as well as telecommunications, wireless, and biotechnical companies. All are bound together by dreams of omnipotence.

In an area of little more than fifty miles (from Santa Clara to Marin County where Pat and I live), a unique combination of products and services has been created that coalesced to make start-up companies develop quickly. It has been said that the Silicon Valley area has sufficient skills, products, and services to enable any entrepreneur to build a company in a week's time! The very fact that this can occur may be one of the reasons so many start-ups have disintegrated—also in a short time.

Silicon Valley has now become a generic term for the high-tech industry. First of all, it has extended its range from the valley in which it originated to forty-five miles north to San Francisco, and an additional fifteen miles west to the city of San Rafael. Second, it is a term the media uses to include Microsoft and Amazon both of which are in Seattle, Washington, and Dell Computers in Austin, Texas. New York City mimics the area by calling its own high-tech industry "Silicon Alley." Consequently, when you see the term Silicon Valley in the media where you live, it usually will signify a corporation or industrial complex that is interchangeable with "high-tech" wherever it is located. Attempts to duplicate Silicon Valley exist throughout the world. For example, Malaysia is trying to create its own version and is calling it "Cyberyaya." To avoid any confusion as to the location of a high-tech firm or industrial complex, in this book we will be identifying Silicon Valley as a northern California entity, and will use "high-tech" for all other areas in the world that are trying to become clones of Silicon Valley.

The effect on the quality of life of the tens of thousands of men and women who migrated to Silicon Valley and who believed fortunes were to be made by everyone overnight during the gold rush years of the mid-1990s is the theme of our next chapter.

SLIDING DOWN THE SLIPPERY SLOPE OF UNEMPLOYMENT

The head of the company in June 2001 fired seven thousand people. He was distraught and guilt-ridden for doing so, and he sponsored a job fair and told the fired workers present: "My deepest apologies to you and your family for laying you off. I am determined to find you a job. Please forgive me." He told the press: "I need to bring hope to everyone."

Is this the new economy, the economy that would change the world? After all, Silicon Valley chief executive officers (CEOs) bragged about the new relationship between workers and bosses during the high-flying decade of the 1990s. Workers and bosses? Those were twentieth-century phrases, according to the CEOs. They were outmoded, old-fashioned, irrelevant concepts. Their new mantra was: "We're all equal, we're all a team, we all work as owners of the company because we all have stock options. So unions are passé, since we *all* are owners who have a common interest: the harder you work, the more the company will grow, and the more money you will make. Unemployment is a meaningless word in our new economy. Our growth is permanent, so there is never any reason for company layoffs. Place your faith in the company, which will never let you down."

This sounded very reasonable to the young high-tech engineers, programmers, and human resource functionaries so many Valley companies employed. The tens of thousands of men and women in their twenties, fresh from major universities with an MBA or Ph.D. knew little about employment practices in the real world, since the Valley jobs were usually one of the first in their careers. The Valley companies made a specialty of hiring new or recent college graduates rather than seasoned workers in the high-tech industry. These young people were naive, malleable, and could be employed at a cut-rate wage scale. And because most of these eager young people were unattached, they had no wife or children to prevent them from dedicating their entire waking life to the company that employed them. The new world left no room for unemployment, but should that happen, since no prediction could be 100 percent perfect, the company would still take care of them.

Alas, this euphoric dream (for that's all it was) was the equivalent of an ecstasy drug experience. The owner of the world-renowned company we quoted did not live in Silicon Valley. He was Lee Jang Dae, chairman of Daewood Motor Company in South Korea, the equivalent in Asia of General Motors. His compassion was so abnormal in business practice that it made front-page news in the June 26, 2001, issue of the *Wall Street Journal.* The *Journal* also reported that only 601 of his former employees found work in other companies. So he, too, was another industry icon with feet of clay.

LEARNING THE BRUTE FACTS OF INDUSTRY LIFE

The high-tech workers who have been "downsized"—which is supposed to be a kinder word than "fired"—learned that the corporate bottom line of Silicon Valley companies was "the bottom line" and that the word "compassion" only existed in the dictionary, not in company practice when economic disaster hit the high-tech industry soon after the new millennium began. Of course, many Silicon Valley high-tech engineers, programmers, and designers were accustomed to working on short-term projects and being laid off once the projects were completed. They then would be employed by another and another company on new projects.

So in that sense during the 1990s short-term employment

was considered "normal" since it really wasn't "unemploy-ment." You could always get another project, another job, at a better salary. In the 1990s, Silicon Valley was a cornucopia of new jobs and new start-ups. "Growth" (another Silicon Valley mantra!) seemed endless.

But the new millennium turned out to be a different world. "Growth" had meant promises of wealth and permanent employment. As John, a skilled programmer told us: "Before last year, I couldn't walk down the street without being mugged by a job recruiter for a company that was competing with the one I worked in! I felt loyalty to the company I worked in—it was continually growing and I was growing with it. A hundred thousand a year and stock options. I was only twenty-seven and felt I was on top of the world."

That was in 1999. Today John's former employer is bank-rupt, and his stock options are worthless. He was led to believe that the growth would never end, but it turned out to be a nightmare instead of a glorious future. Growth can be un-healthy: it can be excessive and metastasize as in a cancer; it can pollute the environment; it can destroy a community.

Yes, the companies in Silicon Valley kept obsessing about growth (every quarter the stock had to rise to please the investors). The problem that few saw and others discounted was that the very growth corporations worshiped created a glut of overproduction. There simply weren't enough new customers (businesses and average consumers) who could purchase the increasingly easier to manufacture software, hardware, cell phones, pagers, and personal computers. Where there was once instant demand for anything Silicon Valley created, a glut of unsold products and services that were daily becoming out-moded had to be eliminated from inventories. The stock market had tumbled in March 2000 and five trillion dollars of market value went up in smoke. Spin doctors were calling this economic condition "a bubble." But that bubble operated with the force of a ballistic missile as it destroyed thousands of jobs and literally hundreds of companies that had the life span of a firefly. The venture capitalists who in the 1990s prided themselves on their financial success in shoring up start-up companies and reaping phenomenal profits as a consequence now began to view any new start-up suggestions as the product of people spreading a disease.

While I am writing this, I am looking at two articles that have just appeared in my hometown newspaper, the *San Francisco Chronicle*. Ironically, the words of this paragraph are being written on July 4, 2001. It is ironic because the Independence Day for thousands of Silicon Valley employees is the unwanted "independence" of not working and not being paid when they want to work. For one article's headline reads "Dot-Com Carnage Continues." It goes on to say:

> The Internet industry suffered another bloody month in June as 53 companies closed or declared bankruptcy, pushing the casualty count 46 percent higher for the first half of this year than all of 2000.
>
> So far, 330 Internet companies have failed in 2001 compared with 225 last year, according to Webmergers, a San Francisco research and advisory service.
>
> The total for June is slightly lower than the 60 tallied in May.

But the data is in line with every month since November, when dot-com casualties first reached 40 and then never declined. . . .

Among the companies closing or declaring bankruptcy in June were PSINet, an Internet service provider in Ashburn, Va., and iTango Software, a human resources software company in Seattle. There was also Bluestar Communications, a regional Internet service provider owned by telecommunications firm Covad, and VitaGo, a health products retailer in Germany.

The other article's headline stated: "Silicon Valley becomes a ghost town for a week." It continues:

The vast headquarters of Sun Microsystems, the computer server manufacturing titan in Palo Alto, was as deserted as a high school during summer break yesterday.

The normally busy parking lot, big enough for the needs of a small city, was empty except for a few cars. Only a handful of people were at their desks, mostly in case of an emergency.

Wounded by a slowing economy, technology companies across Silicon Valley have taken the unusual step of shutting down this week. In an effort to cut costs, they have asked or ordered thousands of workers to go on vacation, creating an eerie calm in offices where frenetic activity is the tradition.

Among the companies using the vacation as a part of their survival strategy are Adobe Systems, Xilinx, Network Appliance and Applied Materials. They hope to save money by turning off the lights, reducing the number of contract workers and by eating into employee vacation time, considered a debt on corporate balance sheets. . . .

Brent, a twenty-five-year-old with an MBA from the University of Chicago, told us, "I'm really worried. The company made it clear I *must* take the week off. I had no choice. But does a week off mean a permanent layoff soon? Nobody tells us anything. It's all so secretive. This is not what I bargained for when I left Chicago to work in Silicon Valley."

Coming from a comfortable middle-class home (his father was a pediatrician, his mother was a dress designer), Brent had never experienced unemployment in his life. For him "depression" was a medical term, not a description of unemployment.

Brent should consider himself one of the luckier ones. After all, he's still on a payroll. What about the "Dot-com Carnage" that the *San Francisco Chronicle* noted? The dictionary defines "carnage" as "massive slaughter as in war." We have often heard from our dot-com clients how their companies were fiercely competitive, using the phrase, "We're at war with the competition." And in that war there are winners and losers; to win, "killing" the competition was necessary.

When sales and earnings dropped, the companies became "lean and mean," as media financial commentators phrased it. "Lean and mean" was just another way of saying sizable numbers of workers had to be fired to cut costs and keep the corporation's stock-market price healthy. It is a fact of financial life that when a corporation downsizes its workforce, its stock-market price usually rises. This makes the investors happy and the workforce sad. The employees become the first casualty of the business war for survival.

Of course, layoffs are only a short-term solution. For if many workers in too many industries nationally are laid off, a recession or depression can occur. Prosperity exists only when consumers have sufficient purchasing power (two-thirds of our national income is based on consumer purchasing power). Since working people are the major consumers in our society, they power our economy; when working people are laid-off for a considerable length of time, they weaken the economy. Unemployed dot-com workers may still have grandfathers who remember the 1929 depression that lasted ten years and was only turned around when massive government spending was undertaken to prepare the nation to enter World War II. Of course, the year 2002 is not 1929, but can we trust the pundits who proclaimed that the new high-tech economy guaranteed prosperity forever before the current business downslide to be accurate when they now say prosperity is just around the corner?

EVERYTHING OLD
IS NEW AGAIN

In the last chapter, we noted that high-tech-company recruiters never mentioned the possibility of layoffs or long-term unemployment when they sang their siren-song to new college graduates back in 1995. The new Silicon Valley recruits were to be given exciting projects, teamwork, loyalty to getting the job done, and the yellow brick road of stock options as the payoff for eighty-hour work weeks. It wasn't a job they offered; it sounded more like a permanent paid holiday.

And why not believe the recruiter's pitch? After all, weren't these new graduates the cream-of-the-crop, the brightest, most inventive, passionate lovers of the new technologies that flourished in endless variations that they could build their careers on? Here was a once-in-a-lifetime chance to be in-at-the-creation of a new world—and maybe become a billionaire, too!

When they came to Silicon Valley, the eighty-hour weeks didn't phase them. It was a small price to pay for the "success" they saw at the end of the tunnel. But what they *now* realized was that the bright light of success they saw at the end of that tunnel of sacrifice (work performance was all that mattered!) was the bright light of an oncoming train rolling over them—the train of unemployment. They had even bragged about working eighty or more hours a week, as if it was a badge of honor that separated them from the average person who "only worked" forty hours a week. They would smile in agreement when they read the Lexus ad that shouted, "Sure we take vacations. They're called lunch breaks."

Before the March 2000 slump, Silicon Valley employees acted on the assumption that any corporate layoff simply meant time to relax, getting another job at one's own convenience at another business, or returning to the old place of work a short time later when a new project began. It was usual practice, no big deal, no stain on one's record. But that was then and this is now, almost two years later. From our research observations, Valley employees now were sliding down a strange new slippery slope of what was beginning to look like permanent unemployment with little or no hope of returning to their previous employed status in a few weeks or months. Here's the timeline of their psychological slide down the slippery slope that started with their job termination. (They are the 25- to 30-year-olds who were the high-flying producers in the industry.):

- **DENIAL**: "Been here before. . . . It's only the usual layoff. . . . I'll be back in a month. . . . In the meantime, I'll have some fun at some of these new pink-slip parties. . . ."

- **ANXIETY**: "It's been a month now and more of my friends are being laid off. . . . But it'll turn around soon. . . ."

- **HOPE**: "No need to worry; this is just a small economic bubble like the papers say. . . . Every company will return to normal soon. . . . Besides, the industry can't do without me. . . . My skills are too great, I'm really needed! . . ."

- **FRUSTRATION**: "Two months off and my bank account is starting to leak. . . . What's happening? . . . Why is it taking so long to get better? . . ."

- **ANGER**: "It's four months now, and I've been turned down by a dozen companies that once would have employed me in a minute. Doesn't my training, my Ph.D. mean anything any more?"

- **RAGE**: "I was told my job would always be there for me and that I could become a millionaire. What bullshit! The stock's now worth 25¢ a share, my options amount to zero, and I might even have to pay taxes on them although I never earned a cent! The company really fucked me over. The CEO and his buddies took the cream and left the rest of us with mud in our coffee!"

- **DEFIANCE**: "I'll show the fuckers. I'll get a better job; I'll start a company and give that old prick of a CEO the finger."

- **DEPRESSION**: "It's now nine months and no sign of any recovery. I'm only kidding myself. Even with all my good ideas, I couldn't start a company—the venture capitalist money's all disappeared. . . . The pink-slip parties used to be fun nine months ago, but God, they're so depressing now. I hope I don't look as worried as most of the guys and girls I saw at the last party."

- **ACCEPTANCE**: "Maybe I better cut my losses. By the time the Valley gets better, I'll be an old man. I've seen a lot of guys looking worthless and trapped because they have a family to support and don't know how much longer they can hold out. Lucky I'm unattached—I'm thinking of moving to another state. . . . Who the hell would want to work in California when you can't even count on enough electric power. It's like a third world country!"

Each psychological way station on this slippery slope of unemployment registers a diminution of personal bravado: the longer unemployment lasts . . . the more rejections encountered . . . the companies without number that tell you you're "overqualified" . . . the networking that turns into pink-slip parties of mutual misery . . . the headhunters who assure you of a job maybe tomorrow, but certainly not today.

Fear starts to eat at the soul, and becomes more ravenous when it appears the job (any job!) in your chosen field may never be found and many will take this as a sign of personal failure ("I was in the top 1 percent of my college class and was certain I would be a millionaire by thirty. Instead I'm broke and I don't know what to tell my parents who expected so much of me. It's horrible, I'm washed up at twenty-eight!"). And there are the others who convince themselves after a year or so of unemployment that it's still only a temporary stumble on their road to ultimate millionairedom or billionairedom. ("It's just a learning experience; hell I'm no loser," we've heard quite a few say. "I'll start a new company with my friends and we'll tell our former CEO to fuck himself!") This false bravado is designed to reassure the unemployed workers that the dream of money and power hasn't vanished in the horizon. The louder they voice this hope for their future ("Well maybe it'll be forty instead of thirty that I can retire.") the more it sounds like a Hansel or Gretel wandering blindly in the woods calling out for a rescuer.

Pat and I have interviewed literally hundreds of men and women who have lost their jobs either recently or a year ago and still haven't found work in their chosen fields, and we have worked with many as counselors trying to help them overcome depression, drug addiction, a sense of failure and terrible loss, self-isolation, uncontrollable anger, rage, and divorce that were the consequences of this unprecedented turn of events in the high-tech industry. We will address in detail in a subsequent chapter the psychological affects on these men and women ITs (information technologists).

Here we limit ourselves to reporting the ways in which literally hundreds of thousands of IT workers have responded to what they thought would never happen to them (to others, yes, but weren't they the chosen few?). Because the high-tech industry has a worldwide influence of enormous significance, what we have discovered in our own research has been confirmed repeatedly over the Internet and in all other media, not only in the United States, but throughout the world. Here are some of the blunt reports (sans media-hype) that illuminate the varied forms of disarray these ITs are expressing:

As if to confirm what we have been saying about the personal shock that ITs are no longer invulnerable to long-term unemployment, an article has just appeared in *Fortune* magazine, dated July 23, 2001, headlined: "White Collar Blues. Free agency is over. Layoffs are back. Many of the people losing their jobs are white-collar and college educated. You could be next."

The article, written by Betsy Morris, reported that the new trendsetter in industry is "smart, college-educated, upper middle-class and unemployed. . . . Since January, announcement of job cuts have come rapid-fire: 26,000 at Daimler Chrysler; 16,000 at Lucent; 9,500 at PG&E; 8,500 at Cisco; 8,400 at Sara Lee; 4,000 at Disney. The list is long and getting longer; some companies, including Nortel, Dell, and 3Com, recently announced a second round of cutbacks only a few months after their first. Hardly any company is immune. John Challenger, CEO of the outplacement firm Challenger Gray & Christmas, calls the carnage 'instant, no-fault job loss in the old manufacturing economy; blue-collar unemployment always rose and fell in lock step with factory inventories; now a similar thing is happening to the mostly white-collar workers in the sleek offices of the new economy.' Morgan Stanley estimates that 81% of layoffs in March and April were white collar."

As far back as 1994, a *Fortune* cover story called "The New Deal," predicted the end of "the widespread replacement of the job contract past of the previous era, the one that traded loyalty for job security. That deal is virtually dead. . . ."

The July 3, 2001, article ends with this prediction of Stephen Roach, Morgan Stanley's chief economist: "In a climate like this, layoffs will continue to rise." He expects unemployment to approach 6 percent by the end of next year. It's very different from the early 1990s recession, he says, and "not a situation that will turn on a dime. . . ."

Jason, a Web designer we know, has been standing in an unemployment office line for so long (eight months since he was terminated) he says he's made friends with fellow IT's who were standing in front and back of him, all worried like he is. "I'm very scared," he told us. "Every time I was laid off before, it was because my contract ended and a month later I'd be in another project. Now there's no more projects. What in God's name is happening to this economy?"

Jason sounded to us as if he just stepped out of the *Fortune* article we noted above.

There was a time when being laid off was considered to be a dirty little secret by the person whose job was terminated. In the past, we counseled men who pretended to go to work every day after they had been laid off so that their wife and children would not regard them as "failures" or "losers." These were the labels they imposed on themselves. Of course, when their "dirty little secret" was exposed, their family was supportive rather than condemnatory.

But that was two or more decades ago; they are parents, or

grandparents in some cases, of the dot-commers of today. Today, however, unemployment comes with the territory of working for any company. It's not considered shameful, but rather a typical risk of the working world. And besides, one can always get another short-term project job whenever one chooses to go back to work.

That was the thinking of workers in the information technologies industry until the industry began to crash in the spring of 2000. Now the feelings of personal loss, betrayal, confusion, frustration, rage, and cynicism have begun to emerge as unemployment appears as a long-term possibility rather than a short-term interruption in one's chosen career path.

Because their attitude toward unemployment is so different from past generations, dot-commers freely express their discontent and disarray in the face of their long-term unemployment experience. The Internet has been the new, effective outlet for their bitterness and anger at their economic plight. The Internet is the new international venue where dot-commers freely and openly announce to the world their attitude of "I've been wronged! I'm the greatest, so it's outrageous for the companies I've worked for to terminate me permanently!"

There are enormous numbers of information technology workers using Web sites such as NetSlaves.com and Fucked-Company.com as platforms for their outrage. These Web sites, originated by dot-commers themselves, have been well described by Michelle Goldberg in an article entitled "Fear and Gloating," which appeared on Alternet.org on July 17, 2000. Here is what it said: ". . . A group of websites are channeling their anxiety and bile into venomous but oddly vibrant communities. They're the dot-com death watchers, and their swelling numbers suggest that a backlash is brewing against the Internet economy. As the dotcom layoffs mount . . . so do the ranks of the disillusioned." They have utilized Web sites "providing a forum for venting against slave-driving start-ups, ludicrous business models and betrayed promises of stock-option riches. The oldest is Netstanes.com, a website and mailing list founded at the end of 1998. . . ."

Equally significant as a site that has become a kind of support group for fired workers is FuckedCompany.com which was created in June 2000 and one year later had four million unique users a month and one million page views a day. This site also serves as an outlet for workers in their thirties and forties as well as the twenty- to thirty-year-olds. "Over and over on FuckedCompany's bulletin boards workers in their 30s and 40s express rage at the arrogant kids who assumed that older people couldn't grasp the industry. The dot-com industry's ageism is one reason that 38-year-old Linda Lanbenheimer (a software engineer) is finding such satisfaction in the current rash of start-up failures. 'Thirty eight is young, except in the valley' she says, 'In the valley I'm over the hill, and I really resent it . . . being treated like I'm a dinosaur is offensive.' "

Here's a typical example of the bitterness and disillusion felt by terminated dot-com workers. It was posted on NetSlaves. com on June 18, 2001, by I. P. Freely (obviously a pseudonym):

> For those of you looking for work in the high tech industry, you know what I mean: you've seen the ads and want to laugh your ass off—flipdog.com, dice.com, monster.com—

none of them do anything for you really except flood your email with newsletters and interview tips and the like. I mean if you're over 25 years of age and don't know how to write a resume or cover letter or even know how to look for a job, then you should just as well check into one of those special institutions for retards 'cause that's what you are."

Millions of people post their resumes on the damn things, and how many of you get one decent phone call that led to some serious work? . . . But still you are looking for a career, something real, with a real paycheck and some real benefits. Fat chance out there! You're lucky if you can get a job that lasts more than a few months and can pay your sky-high rent. . . . Then the bills just keep on coming and that job at the 711 [7-eleven] doesn't look so bad after all, even though you figured you'd never go back to grunt work with a bunch of tattooed pierced teens that say "wow" and "duh" every 30 seconds. . . . People keep telling you to keep your chin up, and all you want to do is knock them senseless, and then there are those other fools who say 'money isn't everything,' and 'it doesn't make you happy.' Well what do they know? Most likely they still live at home and are making just enough to live on.

But you're not a teenager anymore, and you can no longer sit at home and live off your welfare or unemployment or whatever, and those dreams of making it big, getting a Ferrari, or living in a condo with maid service seem so far away. But for now you've given up on those dreams of going to Tahiti or Alaska, because you aren't ever going to even make the plane fare.

Now your life is going down the tubes, and you start thinking maybe I should just give up and settle for a life with a measly paycheck, second-hand clothes and discount sales!

This letter is typical of thousands of others that have appeared on NetSlave.com and FuckedCompany.com. Some of these letters are a short two or three sentences, others long and rambling, not as blunt, and well written as this one, but all expressing cries of pain and bewilderment. For they were millionaires in their vision when they spent eighty hours a week at a dot-com with options that made trips around the world dance in their minds. Now with the dot-com flameout, with their worthless options in one hand and a pink slip in the other, they are just jobless workers instead.

The gamut of their anguish is in the letter we quoted above: in it there is shock, rage, cynicism, frustration, disillusion, bitterness, and most of all fear of the future.

Many letters exhibit feelings of revenge in response to cruel firings. Here's a typical one, written by a programmer who said he turned in his ex-employers for illegal activities when his contract was abruptly terminated:

> We were fired with no warning, the bosses got sweet severance packages and everyone else got a week's pay. While the security guards were marching me and my now unemployed, ex-co-workers out of the building, I figured why not turn them in? I knew they were using hot copies of programs, and anything I could do to cause my former employer pain seemed a great idea to me. (Wired Now 2001).

The plight of IT immigrant workers in Silicon Valley is particularly brutal since they not only experience unemployment, but also the danger, and in many instances the reality of having to return to their homeland. They planned to live permanently in the United States, brought their families, invested in good homes, and now it was all disappearing like the wave of a magician's wand. Paulina Borsevh, who has worked in the IT industry, and is an excellent technology reporter, spells out the plight of these IT immigration workers (you may have heard of them as H–1B—temporary nonimmigrant visa—workers) in her vivid book, *Cyberselfish* (Public Affairs, 2000):

Our heroic high tech CEOs are lobbying like crazy to increase the cap on H–1B visas, which permit immigrants to enter the country and work for six years. The great thing about this is that H–1B workers typically earn less than native-born engineers. What's more, they are rather like indentured servants: They can't move on to other jobs or complain about working conditions, because their visas are tied to the employers who obtained them for them. H-IBs are a great way for start-ups to have employees who can be intimidated into working however long or hard as their execs demand. There's also the tacit threat (or advantage, depending on whether you're siding with management or labor) with H–1B hires that they can be sent back when whatever skill they had is no longer in demand. Just as intriguing, if high tech workers come in on shorter-term B–1 visas, they can be brought over to the United States to be trained at their sponsor companies—and then sent back to a country where the sponsoring corporation can open a much cheaper operation (probably one not burdened by health, safety, environmental regulations, or pesky taxes), staffed with their trained-in-the-USA employees.

Along with the special discrimination against immigrant workers, Silicon Valley companies also single out the men and women who have the audacity of being thirty and over for the type of discrimination known as "ageism." In our country, being young has always been advertised as the only time in life that is truly valuable. But in Silicon Valley being young has been exulted to ridiculous extremes. We have heard more than one dot-commer tell us in counseling sessions that they felt like failures because they never became millionaires by thirty, which seemed to them "old age," a time to retire! It is ironic that the same CEOs who prioritize firing people over thirty are themselves in their forties, fifties, or sixties.

Ageism is an old "friend" of American workers and is always new again. Some grandfathers of dot-com workers will recall a book titled *Life Begins at Forty*, which became a runaway bestseller back in the late 1930s. The title was shocking in its time since everybody then believed that forty was the beginning of old age. Forty was when a person was supposed to be used up and a waste to society.

Before we smile at how foolish this attitude was, take a look at the February 1, 1999, cover of *Fortune* magazine: It shows a man in his forties looking able and healthy, but the identifying title headlines him as "Finished at 40" and then the story goes on to say "Companies used to pay for experience. Now they want potential on the cheap." Just ask veteran marketing executive Mike Bellik who lost his job to a 28 year old." The article inside reports that Ballik "faces a 40% pay cut just as his family's living expenses are at their peak." And the article further reveals that when 773 CEOs were asked at what age they felt people's productivity peaked, the average response was forty-three.

So the "new" trend is the old trend. When we showed the *Fortune* article to a systems engineer who had just been fired (and who knew he was being fired because he was forty-four while his assistant who was twenty-eight continued to be employed) he was furious. "The bastards," he exploded. "I'm more productive now than I've ever been in my life. Yet these sons of bitches consider me as if I'm a piece of toilet paper they can wipe their asses on. If that's their attitude, how in hell am I going to get a decent job in my profession again?"

Another "new" trend has turned out to be an old trend. It's the belief that there are no workers and bosses in the high-tech industry. "You're not a worker, you're an owner" has been the mantra of company executives, and it is used whenever IT workers try to organize into a union to obtain better wages, working conditions, benefits, and job security. In fact, it is the favorite brainwashing sentence that's endlessly repeated by managements such as Amazon when its employees try to better their lot. It sounded reasonable to believe you're an "owner" of the company when you've been given stock options (in exchange for low wages and an eighty-hour week). But now these options are only pieces of paper, not money in the bank. And with options turning into an illusion about becoming wealthy, and with management firing you on a two-hour or one-day notice, circumstances tell you that, yes, you are a "worker" and management is "the boss," and you are vulnerable at all times to be fired if the bottom line turns red. The MBAs or Engineering Ph.D.s are no guarantee that workers are any less vulnerable than the warehouseman when a company wants to fire employees.

We have heard many high-tech men and women we interviewed relate how shocked they were to realize they were not treated like equals in a team-oriented company, but instead were abruptly terminated without notice. We particularly remember Alice, a twenty-five-year-old project manager who had a Ph.D. in computer science, who told us she was stunned when the company she worked for said it was terminating her permanently because the company had filed for bankruptcy (which meant no options or severance pay for her). "I thought being fired like that was what happened in the olden times. The company had told all of us that we were one big happy family. Well, families don't just kick members out the door in times of trouble. No, a company is not a family."

What Alice learned was that some people are more equal than others as George Orwell aptly pointed out. And the most equal of all were the company officials who could fire you. Being fired was another way of recognizing that workers and bosses do exist, since the relationship is indeed based on the power to hire and fire rather than a "we-are-all-in-this-together-so-we-will-get-rich-together" illusion.

History has not been a subject IT workers have majored in. When we inform some of them (a typical group) that the phrase "you're not a worker, you're an owner" had been used by employers throughout the last century, they seem surprised and confused. Workers in the last century were encouraged by very many companies to buy stock in the firm they worked for.

Or if it wasn't stock, it would be bonus payments linked to increased productivity, or profit-sharing with management. All of these tactics were designed to make workers believe that they, too, were "owners" of the company they worked for, that they and their CEO were equals, since everyone was an "owner." Of course, the depression of 1929 shattered this illusion for millions. But after World War II, the illusion returned in full force. Now that Silicon Valley has become noted for its enormous layoffs, there is a pervasive feeling of déjà vu: it's a worker and boss universe after all.

There is a final irony in this situation. The very same high-tech ITs who despised the label of "worker" and who gloried in the Ayn Rand philosophy of I've-got-mine-and-to-hell-with-you, and revile "government interference," show no hesitation about appearing at the unemployment office where they stand in line waiting to obtain their unemployment checks.

They apparently are unaware that the checks they receive are coming from a government-operated program (fought for and won by millions of demonstrating men and women in the 1930s) and that they receive their checks precisely because they are unemployed *workers*.

THE MEDIA HYPE THAT HIDES

THE COLD REALITY OF SILICON VALLEY

"The moral flabbiness born of the exclusive worship of the bitch-goddess SUCCESS. That—with the squalid cash interpretations put on the word 'success'—is our national disease!"
—William James (America's greatest philosopher, letter written to H. G. Wells, 11 September 1906)

There was a time in the long dead days of the years between 1995 and 2000 when Silicon Valley trumpeted itself as the creator along with the Internet of "the largest legal creation of wealth in the history of the planet." And since the spring of 2000, approximately $5 trillion of "wealth" has gone up in smoke!

In a remarkable display of humility (a value in scarce supply in the Valley) the man who originated the above statement (which implied that history only began with the founding of Silicon Valley) apologized for his remarks. His name is John Doerr, once labeled "the visionary venture capitalist" who had backed Amazon and Netscape.

Doerr is such an important player in the Silicon Valley gambling playpen that his apology was headlined in the *Mercury News*, the excellent Valley newspaper. The *Mercury News* headline on July 15, 2001, read: "Top VC Doerr apologizes for helping fuel dot-com frenzy!"

At a meeting of technology industry leaders in San Francisco, he said:

"I'm here today with something of an apology," he said. He then offered a revised version of his quote, which should now say that what had occurred was "the largest legal creation (and evaporation) of wealth in the history of the planet." The article went on to report:

"By billing wealth over innovation, Doerr says, his oft-repeated quote helped fuel a dot-com frenzy that focused more on a 'mercenary drive to make quick fortunes off Internet start-ups than on incubating businesses with revolutionary technology. . . .'

"Doerr also 'bluntly stated that the nation is experiencing a technology recession that has spread to Europe and Asia and could spark a broad-based recession throughout the general economy. . . .' "

Every few months we still read forecasts from "noted experts" that Silicon Valley's economic losses have bottomed out and that the future is bright. But if they would say that "the future lies ahead," they would be on safer ground than their fantasy forecasts. So it is refreshing indeed to get the cold facts instead of the chat hype from an insider like Doerr who has been venerated by dot-coms and ITs throughout the world.

Gloomy forecasts have always been either avoided or disparaged in the land where the mantra of "All-I-need-is-one-lucky-new-break" prevails.

As psychologists, Pat and I are well aware that everyone wants to think the best of himself or herself. To tell a Silicon-Valley high-tech denizen that fear and greed are his or her bottom-line motivation for "making it" in start-ups and stock speculation, would encounter their furious resistance. It is more comforting to oneself to proclaim, "I'm out to change the world" as Steve Jobs, an Apple Computer founder, repeatedly voices. What he has done instead is to change his income, so that for the year 2000 he got the largest income of any CEO in the country, which was $381,000,000. A poll of Silicon Valley and world technology centers reveals that Mr. Jobs is not exactly a beloved individual, nor a generously philanthropic one. They will acknowledge his hard-fisted potential for making a fortune, period. They envy him, and fear him if he is their rival. All other emotions are irrelevant.

Consequently John Doerr's comment that the dot-com frenzy was driven by a "mercenary" drive to make quick start-up fortunes rather than any desire to enhance society's welfare was the equivalent of clearing the air of smog-infested "changing the world" hype. A brave thing to do since, as I've said, people like to think of themselves as public benefactors rather than personal benefactors who derive their fortunes at the expense of the public.

FLASHBACK TO THE DAYS
OF GLORIOUS GREED

The time is 1999. The NASDAQ (the index you see on TV every day that deals in high-tech, SiliconValley style stocks). The media has glamorized the alleged "fact" that over sixty millionaires are created each day through start-up companies and stock options. The general public is steeped in media prostration before these glamorous twenty-five-year-olds who have made more money in one year than practically everybody in this universe makes in a lifetime.

College men and women are electing to get MBA or computer sciences degrees at the expense of "old fashioned" occupations like teaching or psychology or medicine. Money and the promise of the "freedom" it brings captivates their souls. Typical of its time is an article that appeared in the Op-Ed section of the *New York Times* on October 8, 1999. It is written by T. H. Watkins, a professor of Western American studies at Montana State University. He indicates how the SiliconValley virus of instant wealth is all-pervasive among the young students throughout the country including places like Montana. They were called, in 1999, the "boom generation":

> Limitless prosperity, endless economic growth—such is the mantra that echoes in the halls of the Republic.
>
> The resulting mood is pervasive. Here at Montana State University, I teach men and women between the ages of 18 and 25, for the most part. This is not a school of rich kids. Many have to take on one or more part-time jobs to afford the tuition even a land-grant college must charge.

Yet they are sleek with optimism, as certain of the future as if they had all been born with silver spoons in their mouths. During the summers, they somehow get to places like Nepal and Kenya, Mongolia and Cambodia. Many talk of taking a year or two off after college to see even more of the world before turning to the productive work that surely will await them.

They save little, spend much, use their credit cards as if they were rubbing enchanted lamps. And why shouldn't they? Everything they know, everything they hear and see, from MTV to the financial pages, when they read them, tells them that tomorrow is golden, that the great middle-class dream of affluence that has driven so much of our history is finally and inevitably theirs for the taking. . . .

Many pundits say there seems to be no reason why this Sara band of prosperity cannot continue indefinitely, and confidence continues to ring from nearly every quarter of society.

That includes the young men and women who surround me. I look upon them and wonder if my generation is the last to remember that there is no such thing as limitless prosperity. . . .

What Professor Watkins so well described sounds like ancient history today in 2001. Two years later is the equivalent of a lifetime as we measure the economic changes in American life. The "boom generation" is now suffering from a gigantic loss-of-illusions hangover. Suddenly the "old" professions are looking attractive; perhaps teaching, psychology, art history, medicine, and so on are worthwhile careers after all.

But the dream of making it, of striking it rich, of a lucky break, of winning the lottery or the jackpot on programs like *Who Wants to Be a Millionaire* is still pervasive. Of course, to say you are going to start a dot-com today is to invite derisive laughter (over five hundred dot-coms flamed out of existence in the first six months of 2001). Young college students today are witnessing former twenty-five- to thirty-year-old dot-com specialists, who once considered themselves kings of Silicon Valley, returning—sheepishly or reluctantly or defiantly—to old-time jobs now that their belief in wealth beyond one's wildest dreams has shattered. So it's been back to the law office or the reporter's clerk or the bartender's station or the clerk in the neighborhood supermarket. Many of them avoid saying they were in a start-up dot-com (which earned them enormous admiration in 1999, but today is a sign of their economic failure). They prefer to identify themselves as information technologists who are temporarily between jobs in their chosen field. These are the people who admired CEOs like Charles Schwab, founder of the giant investment firm, who nurtured their belief in the wisdom of statements like the following which he wrote and permitted to be incorporated in the book, *Money Talks*, edited by Robert W. Kent, Facts on File Publications, 1985. It is listed in the book as one of "the 2500 greatest business quotes":

> I have gone through some rather dark chapters in American industrial history. It is a great joy to realize that humanity rules today; that industry has awakened to the fact that the employer, in engaging men's services, is entitled to use them

but not to abuse them. . . . Let us hope that the new order which we find in industry will hasten the day when we shall cease altogether to talk about a separation between labor and capital and begin to think of ourselves as contributing to a cooperative undertaking in the advancement of which every supervisor and every employee is an important and essential factor.

Charles Schwab

But in his massive San Francisco–based organization, he has already fired thousands of workers as of July 2001 and may fire more if business continues to decrease. The firings have nothing to do with the ability of the people being let go (research indicates that in an economic downturn only 5 percent of downsized workers are fired for inefficiency). Mr. Schwab, like so many other CEOs, may have the best of intentions, but the bottom line and his own personal economic situation are the ultimate determiners of who works and who doesn't.

THE ICONOGRAPHY OF MONEY

In the June 25, 2001 issue of *Fortune* magazine an article with the following headline is featured: "He's brilliant. He's swaggering. And he may soon be genomics first billionaire."

"Brilliant . . . swaggering," when we hear these words, images of Robert Redford or John Wayne may dance in our minds. But then turn to a full page picture to the left of the article and what we see is a rather pleasant-looking man in his late fifties or early sixties, looking rather self-consciously toward the photographer. He has a tight little smile and the glasses of a lawyer. He wears a conservative brown suit, white shirt and red tie. His hands are rigidly close to his body. He certainly doesn't fit the image of those two dazzling words. Of course he may very well be an outstanding CEO, which he indeed is— he is Bill Haseltine, CEO of Human Genome Services. *Fortune* magazine, however, glamorizes him as "swaggering" precisely because he may soon be genomic's first billionaire. The impression given by the headline is that if you are likely to a billionaire, you have all the outstanding qualities of being a great human being. Before the March 2000 dot-com bust, CEOs who became billionaires were praised as if they were rock stars: world-famous because of their fortunes, to be admired and envied as the heroes everyone wanted to become. The lust for giant fortunes was equivalent to the pursuit of sainthood.

The last sentence is no exaggeration as American society has been imprinted with the pursuit of money as the precursor to beatification. For the favorite buzzword the media—all media!—invariably uses to describe the CEOs of the high-tech industry is "icon." An "icon," according to the dictionary, is "one who is the object of great attention and devotion; an idol." It is also described as "a representation or picture of a sacred or sanctified Christian personage."

The names of numerous Silicon Valley CEOs are referred to by the media as icons: Steve Jobs, a founder of Apple Computers, who "earned" that huge $381 million pay package in the year 2000, and Michael Dell, founder of Dell Computers, with a $94 million pay package in the same year. Then there are a host of other high-tech CEOs who have been blessed with the same "icon" honor: billionaires Larry Ellison, founder of Oracle; Andy Grove, a founder of Intel; CEO John Chambers of Cisco Systems; Jerry Sanders, founder of AMD; and, of course, the towering icon of all icons, Bill Gates, founder of Microsoft. If you notice, Gates is now referred to in the media, most of the time, as William H. Gates III, a title appropriate for the man who is the wealthiest in the world. He has moved up the ladder from "call me just plain Bill" to "honor me now as William H. Gates III."

Before the recent economic downturn, these CEO icons were exhibited as people holding positions to which any Silicon Valley mid-level employee could aspire. This media hype was screamed to the world by even responsible publications like *Business Week*. It published a special issue on the "Silicon Saga" on August 25, 1997. One of the articles was titled "Even the Receptionists Are Millionaires." It told the story of two women, one a receptionist, another a personal assistant. They became wealthy through stock options, buying for example stock for $5.00 a share that was worth $2,000 each fourteen years later.

The impression *Business Week* gave was that anyone could become a millionaire in Silicon Valley by working at a low-paying job that would magically transform itself into "multi-million-dollar opportunity."

At the other end of the economic spectrum, *Business Week* in the same issue reported on CEO Samis Arora, founder of NetObjects, Inc., who became a multimillionaire at the age of thirty-one. This article, entitled "My Jet Is Bigger Than Your Jet," then went on to whet the appetite of the reader:

A booming tech market and stock-option culture has helped swell the ranks of Silicon Valley millionaires by 45,000 since 1994, to 186,511 in 1996, according to Payment Systems, Inc. The area now boasts one of the highest per-capita income levels in the country—$32,548 in 1996 for the San Francisco Bay area, vs. a national average of $24,324.

In the Valley's toniest neighborhoods, one might think money really does grow on trees. In Woodside, for example, where high-tech luminaries such as Larry Sonsini and Scott Cook rub elbows with neighbors Neil Young and Riley Bechtel, "starter" homes begin at $800,000. Most sell before they even hit the market, and buyers often plunk down cash, as Arora did for a sprawling $2 million home in the Woodside hills.

Not far away, Oracle CEO Larry Ellison is building a $40 million-plus replica of the Japanese Katsura palace on 23 wooded acres. Presumably, he'll be parking his new Marchetti Italian jet fighter elsewhere—perhaps next to the aerobatic plane he just bought as an eighth-grade graduation gift for his son. In Atherton, Tom Proulx, a co-founder of Intuit, recently bought three neighboring lots so he could build a 9-hole golf course in his backyard. One Woodside programmer bought 24 acres just to land his helicopter. "Keeping up with the neighbors in Silicon Valley is getting weird," says Daniel Case III, president and CEO of Hambrecht & Quist, Inc., It's not, "do your kids go to private school." It's "do they have a private jet?"

After reading articles like these, it was no wonder that many people who came to see Pat and me for help in solving their interpersonal relationship problems expressed guilt and anxiety, bemoaning the fact that here they were thirty years old and were personal failures. They were employed in career-level jobs in the media and business, making better-than-average salaries, but were nowhere near being millionaires. The media hype made it seem so easy to become one in 1997, so the belief was perpetuated that it *must* be a sign of your personal failure if you didn't have a million dollars in the bank. Retirement at thirty was the goal of the Silicon Valley millionaires, so many people felt apologetic that they still had to work for a living.

This attitude that the media generalized seemed perfectly reasonable in late 1995, but only because the media focused on the so-called winners of the millionaire lottery, when all they represented was a minute fraction of the majority of people who did the nonglamorous work in Silicon Valley. A myth was perpetuated and it still lives today, but in diminished form. In 2001, the "Icons of Silicon Valley" are still called "icons": they were the men who, for the most part, saw their opportunity in the 1970s and 1980s and seized that opportunity first. They indeed had a talent for making money. These icons are now in their late forties or fifties or sixties. It is mostly the new generation of later twenty-year-olds who created the dot-coms between 1995 and 2000 that have created and destroyed multimillion-dollar start-up companies in the space of two or three years.

Ever since our country was founded, Americans have had a love-hate relationship with people who have monopolized fabulous wealth. The Declaration of Independence affirmed the right of everyone to life, liberty, and the pursuit of happiness. It did *not* affirm the right to the pursuit of money as a fundamental need of human beings. Of course, subsequently, many, if not most, people began to believe that the pursuit of happiness meant the pursuit of making a fortune. Attitudes such as these have waxed and waned. The great economist Thorstein Veblen created the phrase "conspicuous consumption" to describe the gross tycoons of his time in the early part of the twentieth century. That was the time when multimillionaires and billionaires were not identified as icons in the media. They were called "tycoons," which the dictionary describes as "a wealthy and powerful businessman or industrial magnate." The term "magnate" was also frequently used. It meant "a powerful or influential man, especially in business or industry." And even today you might occasionally see a person described as a "mogul," which is a very rich or powerful person." The word "icon" as it related to businessmen was never used until recently. The old words—tycoon, magnate, mogul—all were identifications of rich and powerful people. There was never an indication in them that these rich and powerful people were to be idolized as icons. In fact, they were frequently reviled as oppressors and exploiters of people: J. P. Morgan, Andrew Carnegie, and John D. Rockefeller were considered cruel, selfish, and exploitive individuals as they monopolized their industries (later of course, with the help of superb public relations people, they were considerably whitewashed).

Money equals power equals freedom. It is this aspect of the American dream that still prevails today, even among the dot-com entrepreneurs whose start-ups are now dreams that disappeared.

Indeed, on the surface, money equals power equals freedom

is a very appealing equation. Michael Wolff is frank about this equation. He is a journalist who started a dot-com that rose and fell in the late 1990s. He wrote the best-selling book about his experience of becoming a dot-com entrepreneur who was defeated by circumstances inherent in the dot-com frenzy, which people thought was an opportunity rather than the disaster it became. His book is titled *Burn Rate* (Touchstone 1998), and here is what he says about his own lust for success:

He was absorbed with "the prospect of making millions quick. Of making more money than you ever dreamed of. Of making the kind of money that would allow you to do all the things you ever dreamed of doing without the bastards getting you down. Fuck-you money. The sweetest lucre."

He was also speaking for a whole generation. Pat and I have heard the words "fuck-you money" so many times in our interviews that it became a tiresome cliché ringing in our ears. And in the very last sentence of his book the infatuation with fuck-you money still persists after his dot-com failure. The sentence reads: "And here I am making plans again."

You have perhaps noticed we have said very little about the women in Silicon Valley's high-tech industry. We think they have been so overlooked, or deliberately neglected by the media that we will be devoting all of chapter 4 to them.

As for today's "icons" of industry, they will appear again in our subsequent chapter on "The Quality of Personal Life in Silicon Valley." At this point it is sufficient to state from our personal experience that these men (and it is an exclusive men's club of billionaires) as persons are neither better nor worse than the average American. They have average looks (their money hasn't made them handsome). However, in one key respect, "the rich are different from you and me," as F. Scott Fitzgerald famously remarked, the difference is that they actually live the equation of money equals power equals freedom. However, it is the size of their fortune, not their ethics or morality, that separates them from you and me. And the power they derive from their fortunes is power that, in Lord Acton's words, is the power that corrupts, and absolute power corrupts absolutely. It is this power coupled with freedom which leads to self-indulgence, self-glorification, arrogance, and the self-deceptive view that indeed they can control the universe. In the last analysis, the wealthy can be called, in Chaucer's words, "smilers with knives," otherwise they couldn't have become the men-on-top-of-the-competitive-heap, or remain on top since others would try to daily destroy their place in the high-tech pantheon.

Since the super rich had the ability to determine their own salaries, it takes no sightseer to predict that the companies they ran were kinder to them than they were to the many thousands in their workforce who were unceremoniously fired during the year 2000 business meltdown in the high-tech industry. The *Mercury News* on April 17, 2001, headlined an article: "In the world of huge CEO salaries, failure doesn't matter." Last year, while profits lagged and typical investors saw their portfolios drop 12 percent in value, CEOs at top firms got an average 22 percent raise in salary and bonus. . . . CEOs who cheerlead for market forces wouldn't think of having them actually applied to their own pay packages. . . ."

In case you aspire to becoming a billionaire, the odds against becoming one are rather steep: *Forbes* magazine, in its year 2001 count, reveals that worldwide the billionaires club consists of 538 members.

THE COLD REALITY
OF THE OLD NEW ECONOMY

The media hype, which takes its cue from the top Silicon Valley spokespeople, has called the Internet and World Wide Web inventions "revolutionary." It would, they have said repeatedly since 1995, change the world, making it a better place to live in. It was a communication marvel that would bring the peoples of the world together and create a universal community of peace and prosperity. It is well to remember that President Bill Clinton's administration took up that belief and urged a computer in every family household as the answer to poverty. Happiness was to exude from a PC or laptop in every household, school, and business in the United States. And as a byproduct of this paradise, recessions, depressions, and long-term unemployment would be banished forever. People would smile at the ignoranance of older generations who allowed a depression like the one in 1929 to cause one out of every four men and women who were willing and able to work to become unemployed for many months or even years. The new economy would be triumphant.

We no longer hear much of this bravado spoken today. Too many similarities to the "old" economy exist at present that cannot be easily dispatched as temporary impedances to the "new economy." First, there is an ongoing international recession taking place. The media calls it "a temporary significant adjustment," "a leveling off," "a restructuring," "the dot-com flu," "a correction," "a bump," "an economic hiccup," "a bubble," "nearing a plateau," "bottoming out," "a restructuring"—these are all terms to convince the average consumer that there is nothing to worry about, so continue to buy, buy, buy, and spend, spend, spend so you can help the economy. But there are thousands of people who have been fired because of lack of business and who can't find decent new jobs; others are forced to take part-time jobs when they want and need full-time employment. And profits and sales in manufacturing and high tech and the service industries are dangerously low, so that even top CEOs have been forced reluctantly to make infrequent gloomy announcements about their current company earnings and future prospects. In other words, it's a recession.

Of course it is different and nowhere near like the 1929 depression, but it does bear a remarkable similarity to the memories of the recessions of the recent past—in the early 1970s and in 1987. There is every indication, however, that the current recession is becoming more severe than those in the recent past.

Recessions in the past were foreshadowed by severe declines in stock-market values. This "old economy" occurrence is very prevalent in the alleged "new economy." In fact the NASDAQ, a Wall Street index that measures the value of high-tech stocks, declined over 50 percent since March 2000. And five trillion dollars of value evaporated. This reads more like a description of the "old economy" than a new one.

Of course, there have been—and will continue to be— remarkable changes in business practices and in our personal lives as a result of the predominance of the high-tech industry. But remarkable changes do not make a brand new economy. The very nature of our capitalist system involves what the famous economist Joseph Schumpeter called "creative destruction." That means new products and new inventions are always prevalent in our society, with the driving force the desire for greater and greater profits. That's the "old economy"—and that central fact underlies the "new economy."

Certainly the question is moot about whether the *new* revolution of high technology is more unique and more transformative than the economic and social revolution brought about by the invention of the automobile. John Keats, in his 1968 book called *The Insolent Chariots*, accurately remarked: "The automobile changed our dress, manners, social customs, vacation habits, the shape of our cities, consumer purchasing patterns, common tastes and positions in intercourse."

The jury is out as to whether or not the high-tech industry will have a similar or more important influence in our society.

There is a wonderful line in the great 1963 movie *The Leopard*, which is stated by the protagonist: "Everything must change in order for everything to remain the same."

What the character meant was that for power to rest in the hands of the people who have it, they must be flexible enough to welcome innovation. Certainly this is the underground stream in American society, where the Fortune 500 Companies have held the power in our society under the "old economy." That power still exists after 1995, the twentieth century employer-worker relationship remains the same today as then. The "free agent," "contract worker," "temporary worker," "contingent worker" of today amounts to 40 percent of all Silicon Valley workers. They are the workers with no security, who can be fired at will, who have no overtime pay and few if any benefits such as 401ks or health insurance. And they are paid much less than full-time workers for the same work. This sounds like the unprotected workers of the early twentieth century. It is certainly an "old economy" concept masquerading under new titles. New titles do not make a "new economy."

A "new economy" was proclaimed by Silicon Valley icons (also called "visionaries") a pollution-free, stress-free, cooperative, healthy-mind-in-a-healthy-body economy. To determine whether this vision is anywhere near reality, it would be best to seek out the recollections of a person who lived in the Valley as a child in the 1960s and 1970s and returned to visit the area in which he had lived, which was Sunnyvale, in the very heart of Silicon Valley just prior to the March 2000 recession. His thoughts were written at the height of economic prosperity, a time when the Silicon Valley high-tech elite bragged that, "First there is Silicon Valley, then there is the rest of the world." He was journalist Jeff Goodell, and he wrote this visceral reaction to the alleged "new economy" in the *New York Times Magazine*, May 28, 2000:

> I should be proud of the stature of my hometown, but the truth is that I have mixed feelings. This whirlwind has been great for my retirement fund (even after the recent battering), and I love my computer almost as much as I love my dog. But outrageous prosperity does not come without a cost: most of my friends and family who do not work for high-tech companies have long since fled the area, driven

out by skyrocketing real estate prices and snarling traffic. My eyes have burned from the thickening smog. I've tasted the metallic water that flows out of the faucet in my old house and I've seen even my most successful, level-headed friend become consumed by envy because the guy in the cubicle beside him is worth nine million and he is worth only four. I've watched the apricot orchards get bulldozed and the dark, rich alluvial soil—the finest in the world, ranchers say—get paved over and forgotten. Twenty years ago, the valley felt like a suburban nowhere, sleepy and homogenous; today, it feels like a giant, nonvirtual eBay, a noisy, chaotic place that is growing at an exponential rate, where everyone is consumed with making the big score. . . .

Jeff Goodell's response to his former hometown could be repeated by any objective observer of what has happened to the entire Silicon Valley complex of cities. The prestigious nonprofit Silicon Valley Toxics Coalition (SVTC) was founded in 1982 as a result of discovering substantial groundwater contamination in Silicon Valley. From that time on, it has drawn attention to the dangers to the environment and employees in the high-tech industry. According to its Web site (www.svtc.org):

Santa Clara County suffers greater environmental and occupational health problems from high technology development than any other region in the world. Yet no public health program exists to track and prevent illness and injuries caused by hazardous chemical exposures. With a rising population of recent immigrants, 1.6 million people make their home in Silicon Valley, the birthplace of the electronics industry. The wealth of Valley executives and affluent engineers contrasts with the growing number of working poor people on the assembly line and in the service sector. People of color, mostly Latino and Asian/Pacific Islander, make up forty-five percent of the general population and fifty-six percent of children under the age of fifteen. Lower income people of color work in the most hazardous jobs and live in the most polluted neighborhoods in Silicon Valley.

Exposure to hazardous chemicals may contribute to known health disparities and increase susceptibility to other hazards, like infection. Ethnic minorities and low-income groups suffer poor health compared to the more affluent, white population in Silicon Valley, losing more years of life per death from heart disease, cancer and cerebrovascular disease (strokes), for example. People of color, especially women, also face greater health and safety risks from industrial pollution both on the job and in the community. A majority of workers in high tech semi-skilled production jobs, which often involve hazardous chemicals handling and exposures, are people of color, mostly women. Yet Latinas and Asian/Pacific Islanders also live in neighborhoods nearest to sites of toxic leaks and spills from industry, resulting in "double exposure" to chemicals, in many cases. . . .

The billionaire-flush high-tech companies have been careful to present to the world a public relations image of themselves as creators of pollution-free, environmentally clean manufacturing sites. Yet their response to the findings of the SVTC charges has been more one of avoidance rather than cooperation to make the valley what its media hype claims it to be.

If there is any car travel you would wish an enemy to drive him or her mad, try a Silicon Valley freeway grid-locked drive. Mora Gunn, a high-tech reporter for the *Mercury News*, in her column of February 15, 2001, remembers driving in the Valley in her youth:

"The Silicon Valley of my youth seemed to be just a bunch of orchards with a single freeway, mostly empty, running down the middle. Today, it looks like a snake ball. . . . For years I commuted from San Francisco and I was able to drive at maximum speed. Today, both directions of the nearly 300 miles of freeway that criss-cross the valley are liable to be stop-and-go."

A natural mobility study by the Texas Transportation Institute revealed that San Jose (the major Silicon Valley city) ranked far higher than even New York and Chicago in road delays. Drivers were stuck in traffic for an annual time of forty-two hours in San Jose and also in San Francisco. That's a waste of almost two days of a person's life. New York and Chicago ranked much lower with wasted time in traffic of thirty-four hours each. As might be expected, Los Angeles ranked first with wasted time of fifty-six hours.

Pat and I have driven to and from San Jose many times and each time we end up looking at each other saying next time maybe water torture would be better.

Rather than creating a stress-free work environment, the Valley consists of a majority of stressed-out workers, accustomed to eighteen work hours seven days a week, and always on call, even at three in the morning (all the skilled high-tech workers have cell phones and pagers, not for personal pleasure, but for them to be instantly accessible any time of the day or night to their company's emergency demands).

The stress overload resulting from sleep deprivation, poor management, and unrealistic deadlines has taken its toll on these employees, the toll of drug abuse. "Drugs are the dirty little secret of the dot-com world." That's the statement of Dr. Alex Stalcup, medical director of the New Leaf Treatment Center in Concord, California. Forty percent of its new patients are from the high-tech industry. "There's so much money, such long hours, such pressure to perform here. It's speed to work on, coke to play on and smoking heroin to come down on," he added.

Dr. Stalcup's findings (supported by the Narcotics Task Force of Silicon Valley's San Mateo County) were reported in the October 4, 2000, issue of the *International Herald Tribune*. Articles such as these, however, rarely received prominence in the media, since the newspapers, magazines, and TV always ignored the dark side of Silicon Valley, because the glamorous side (that is, the "make-your-fortune" side) was much more exciting to emphasize. So the information about drugs, water and air contamination, and traffic gridlock, while occasionally acknowledged, are usually hidden under an avalanche of features about the supposed wild and wonderful craziness of the area.

In Silicon Valley there are many high tech, highly skilled

engineers, Web designers, and human relations administrators, among many other professions, who are never noticed by the media. They exist, but are invisible due to the lack of the publicity they deserve. I am referring to the highly competent women of Silicon Valley, who are the subject of the next chapter.

EXPOSING THE MYTH OF
GENDER EQUALITY
IN SILCON VALLEY

S ilicon Valley's public relations campaign of presenting itself as a "new economy" has trumpeted the assertion that gender equality exists in the high-tech industry, unlike the dead economic past where women were mercilessly exploited and sexually harassed.

Equality of opportunity and job performance, not sexual differentiation, were supposedly the only open doors to success, career advancement, and monetary rewards. In Silicon Valley, you were part of a family in which you would never be unemployed. And your ability and effort to succeed would determine your status in the company, in the world in which you live: we're all friends, we dress informally, we all work in cubicles, we talk to each other as equals.

That was the Wizard-of-Oz story CEOs were fond of presenting to their employees and the world at large. It could best be summarized in a typical corporation Christmas party pep rally speech given by Jerry Sanders III, the CEO of Advanced Micro Devices (AMD), a giant semiconductor and chip technology company that creates the elements that are the heart of personal computers. Here is what he said:

> We're a noble experiment. We're a company that believes in meritocracy. That means the better job you do, the more promotions you get, the more money you make, the more fun you have and the better the company is for you. You're great. I offer my thanks for a wonderful year. You're my family. I love you, Merry Xmas and God bless you.

This was the kind of statement that has been endlessly repeated over the past two decades by countless corporate executives and their media assistants. Jerry Sanders made that statement in 1982, and you can hear its echo today in a recent speech Steven A. Ballmer, CEO of Microsoft, gave to company employees. The *New York Times* of August 13, 2001, reported that a video clip of his corporate rally speech, which has been making the rounds of the Internet, shows him frantically jumping about for almost a full minute saying, "C'mon! Give it up for me. I have four words for you: I-Love-This-Company! Yeah."

It may be harder to think of a high-tech company as "family," or something to "love" today. After all, unemployment today in 2002 has become the norm in the industry, rather than the exception—and that is hardly a reason to "love" the company you worked for. Of course, your family would never give you a pink slip when times are tough. But a company that tells you it is a family can and does.

The René Magritte observation that everything we see hides another thing is certainly applicable to the status of women in the high-tech industry. For hidden underneath all the expansive talk of gender equality and the alleged triumphs of "meritocracy," are the following statistics from the documentary *Secrets of Silicon Valley*, that belie this illusion:

▼ The place of women in the high-tech industry is on the factory floor rather than in higher positions: 95 out of 100 engineers are men. Most of the service workers are women. The majority are undocumented immigrants who come from the third world and live in isolation, confusion, and fear of being deported. Their wage levels are below $10.00 an hour.

▼ The glass ceiling in the high-tech industry is alive and well—it's more like concrete. When Anne M. Mulcahy became chief of Xerox in July 2001, she became only *one of five women* listed as Fortune 500 chief executives. One percent of all Fortune 500 chief executives were women! Perhaps we can consider this progress, since a dozen years ago women were even invisible among the middle ranks of corporate managers, let alone at the CEO level.

DISCRIMINATION AT THE TOP: A WOMAN CEO INSIDER'S VIEW OF THE HIGH-TECH WORLD

In the process of researching this book, Pat and I became acquainted with a remarkable woman, Janese Swanson. She is the exception to the Silicon Valley rule that a high-achieving, successful CEO must be a man who must be driven by the tunnel-vision obsession that winning is everything and that wealth and power are the ultimate aphrodisiacs of life. And to achieve that end, triumph over one's competitors, no matter the cost, is the only thing that matters: ethical qualms are for poor people.

The many men of this type who we had met were so similar in attitude and arrogance that they seemed to become a blur of interchangeable parts. So it was with surprise and delight to discover that a CEO could be a woman who was a warm and open person willing to share with us the complexities of her feelings and achievements in the male-dominated high-tech industry. Janese Swanson is that person.

Janese lives in the same city we do, San Rafael, California, and before we met her, we knew of her outstanding work. Her executive biography lists a record of high-tech achievements, the following is a condensed version:

As a senior product manager at Broderbund Software, Dr. Swanson managed the Sensei educational line and produced the popular Carmen Sandiego™ series. Her Treehouse™ and Playroom™ series received numerous awards including the Parent's Choice Award.

Dr. Swanson founded Girl Tech in 1995 after finding that so many wonderful technology products were developed for and marketed solely to boys. Dr. Swanson's doctoral research on children's play patterns and use of technology revealed (1) that girls are starting to shy away from technology, a factor that will be detrimental to their future career opportunities; (2) that girls enjoy and will play with technology toys/software if they are developed with girls' interests in mind; and (3) it's important to reach girls and help them become comfortable with technology before they hit the confusing teenage years.

Since forming Girl Tech, Dr. Swanson has produced the first of a series of books. *Tech Girl's Internet Adventures*, which won Dr. Toy's 100 Best Products for 1997 and the Parent Council Award, developed a popular Web site (www.girl-tech.com) which has received over 35 awards including CNet's Best of the Web. Dr. Swanson sold Girl Tech in 1998 to Radica Games Limited (NASDAQ—RADA) taking on the role of Vice President of the Girl Tech division until June 2000. Today, Dr. Swanson is CEO of a new start-up venture called eDames, a company dedicated to bringing technology to young women.

In 1999, Girl Tech's first line of seven electronic toys were successfully launched. The Password Journal™, an electronic journal that locks and opens via voice recognition, is a 1999 Dr. Toy winner and one of the top selling electronic toys on the market according to fall 1998 TRSTS. The Password Door Pass™ is a room protector that keeps track of intruders: when the unit detects motion, it requests a verbal password and sounds an alarm. . . .

Dr. Swanson's dedication to helping girls is reflected in

the many awards she has won, including the "Annual Leading Change Award" from Women in Communications, Webgrrls, "Top 25 Women on the Web," YWCA of the USA "Advancement of Girls and Technology," and "Woman Entrepreneur of the Year Nominee" from National Association of Women Business Owners. She was featured in *Ms.* magazine's "Women of the Year" issue, 1997. She was the keynote speaker at the National Inventors Hall of Fame induction 1999 award's ceremony and served on the Advisory Committee of the Lemelson Center for the Study of Invention and Innovation in association with the National Museum of American History and the Smithsonian Institution. Currently, she is on the board of trustees at Dominican University. Dr. Swanson holds seven degrees, including a doctorate in Organization and Leadership as well as an MBA and a Computers in Education Specialist Degree.

Dr. Swanson's credentials and business achievements are indeed very impressive, particularly since they are accomplishments in a male-dominated industry that has little respect or room at the top for women. In terms of male evaluation, Dr. Swanson could be considered "one of the boys," since the high-tech industry considers women like her more of an accident of nature rather than a typical example of what women can achieve.

What makes Janese Swanson very different from the male CEOs in the high-tech industry is the person, the human being, behind the title. For when Pat and I met Janese for the very first time, we were struck by her lack of pretension, by being the exact antithesis of the stereotyped CEOs we had seen. For she was a smiling, dark-haired, forty-three-year-old attractive woman who was tall (five feet, nine inches), slender, and well-groomed. It became quickly evident that our interview with her (the first of many) was not going to be the standard boiler-plate story of self-aggrandizement and "how-I-turned-my-company-around-from-zero-to-a-billion-dollar-business."

Janese wanted to share with us her passionate commitment to educating women and their daughters on the opportunities for becoming high-level achievers in whatever career field they chose to enter. While her chosen field has been the high-tech industry, she has always utilized it as an arena for promoting gender equality and high self-esteem in women in every phase of American life. Other CEOs would routinely say they were in the high-tech industry to "change the world," but all they meant was to make more money and acquire the status money brings them. But Janese actually meant (and means) to change the world by raising women's expectations for themselves: money and status are means to that end, not the end itself.

Another refreshing revelation we experienced was the fact that Janese's rise from rags to riches was true, not a phony pub-licity handout. Where most of the male CEOs came from well-heeled families, and went to elite universities, Janese came from a poverty-stricken family, and she was the first person in that family to ever graduate from high school, let alone college or a university. Her story sheds light not only on the high-tech industry's exploitation of women, but also on the status of women in general in today's society—a status that is replete with mixed signals, double standards, macho intimidation—and also real gains in gender equality combined with continuing traces of second-class citizenship.

Janese can best speak for herself in her own candid words. So, we present below a summary of the many interviews we had with her as she told her own story.

THE JANESE SWANSON STORY

I was born in 1958, so I grew up at the beginning of what has been called the women's revolution. But all I was first concerned about in my early years was sheer survival—and I learned through my surviving that women had a long way to go before society would treat them equally.

I grew up in San Diego, California, and was one of four children—I have an older sister and two younger brothers. My mother was a single mom, since my father went to Vietnam in the sixties, when I was six, and didn't come back. So the burden of supporting our family was placed solely on my mother and what we kids could contribute to survive economically. We had no extended family to help us since my mother came from a poor Latino background.

I have good memories of my father. He was Swedish, so you could say I was a multicultural kid like so many children are today. He was a flight controller, and I remember him taking me into a flight tower where I was surrounded by the technologies there as far as I could see. I can still recall how excited I was—it all looked like I was in a treasure-house in a foreign country. It was so fascinating! I guess you could say my father introduced me to the wonders of technology and that's remained with me ever since.

My father went to Vietnam as an air traffic controller in the sixties, and I used to watch the Vietnam war on television knowing my father was there. Every once in a while he would call us from the field. It felt so thrilling and amazing to me to hear his voice on the telephone because it was from a far away place I saw on TV. He would tell me to say "over" after I finished saying something to him. I was six then and was so happy to hear his voice that I wanted to crawl into that telephone so I could talk to him face to face!

At that time my mother was going to cosmetology school and was working at two jobs so our family could survive. I had to become the caretaker for the family since my older sister was very rebellious. She left home a number of times, while I stayed home and took care of the house and my two brothers while my mother worked.

We moved so much in my teens. In one year, I went to six—yes, that's right, six—different schools. We moved all the time because mom was always running away from bill collectors. We never had enough money to pay the rent. There was this manager of an apartment building where all of us were living in a one-bedroom apartment. He would knock on our door and yell, "I know you're all in there, so come on out!" But we kept silent as church mice—all of us kids were hiding under the bed.

I think I got my first taste of brainwashing because I was a women when I saw on TV the first man in space walk on the moon. I immediately thought, "I want to do that, I want to be the next man to walk on the moon." I was so excited over that fantasy. And then I got depressed, because I had said I wanted to be the next *man* on the moon; but I couldn't because I was a woman. Only men could go to the moon; that was the accepted opinion at the time. But something inside of me, even then, felt that was so unfair.

And when I was eleven, we were living in La Jolla, one of the richest communities in the nation. But I was cleaning the houses of wealthy women rather than enjoying the sunset, in order to bring in a few dollars to our household. At that time I was fascinated with reading books about the human body and decided I wanted to be a doctor. So it felt like an added treat when I went to clean this wealthy woman's house in La Jolla, since her husband was a doctor. Maybe she could tell me more about becoming a doctor, I thought. She was a nice lady who fed me lunch (her elegant, dainty sandwiches still stand out in my mind!). She asked me, "So what do you want to be when you grow up?" I told her I wanted to be a doctor. She said, "Oh, my husband is a doctor—it will be easier for you if you marry one."

I felt hurt by her response. Here I was in the sixth grade, excited about having a career as a doctor, and this lady was telling me I wasn't entitled to become one. She wouldn't have said that if I had been a boy.

The other dream I had at that age was to become a "paper route boy." It was a very modest hope and seemed to me realistic. I was the best bike rider in my community. I would beat all the boys in the races we had. But the newspaper employer told me I couldn't be hired because I was a girl, not a boy. So the title "paper route boy" meant what it said: no girls allowed.

All of this was happening to me in the late sixties and early seventies when the feminist revolution was beginning. But the grip of these old ideas about what women can and cannot do still was dominant. Yet the impact of that revolution was making me begin to think that it was so very unfair that women couldn't have the same opportunities as men. My mother had been entirely brainwashed. For when I expressed my expectations to be a doctor or computer expert, she would say: "That's nice, but wouldn't it be better to be a nurse or beautician, because that's all women are entitled to become."

Because we never had enough money to pay all our bills, I started to work full time while going to high school. When I was fourteen, I lied about my age and could get away with it because I was tall and seemed more mature for my age. I worked at clothing stores and at a fashion store all through high school. My bosses told me I should be a model since I was tall and slim. And for a couple of months I did model but hated it. It made me feel I was only a body, a piece of merchandise rather than a person in my own right.

The job I got the most money for was at Sears during my later teen years. They started me out as a "floater," working in all the different departments. But the one department I ended up at was the record and tape department. I loved music, and I tripled the sales there in my first month on the job. The manager then asked me to do the ordering, and before I knew it he put me on commission sales in the TV, radio, and tape department. I loved the fact that it was technical stuff I was dealing with. I was fifteen and a half at the time (a secret no one knew), and I was the only girl in the department. I was on straight commission with men who were over thirty. The men in my department felt resentful toward me because my sales were far higher than theirs. The guys would tell me that they were the heads of their households with wives and kids to support, so why don't I give them some of the sales I made. Since I always felt guilty because I was making more money than they were, I would give them some of my sales, even though my family needed all the money I made.

I was so young and naive then that I didn't know I was also being sexually harassed at the same time I was giving away some of my sales earnings to those salesmen. I didn't even know what sexual harassment was. It was a subject just beginning to be talked about in the seventies.

What happened was that one of the salesmen used to whistle in my ear and was constantly looking at my legs and my breasts. He was always trying to touch me, and it made me feel threatened and uncomfortable. The salesman was always in my body space, and I didn't know how to deal with that. One day, I couldn't stand it any longer and decided to quit.

The manager didn't want me to quit because I was the top salesperson he had—and he was also a nice guy. So he asked me why I wanted to leave, and I told him it was because the salesman was coming on so strong. The manager then went to the salesman and confronted him about it. He also told the rest of the guys to back off. So I didn't experience any more flack in that department during the rest of the time I worked there, which was from the ages of fifteen and a half to eighteen.

My experience at Sears taught me all men weren't bastards. There are decent men like my manager, and I'm careful not to stereotype all men—but I also feel sad that too many men today in the high-tech field still have that macho contempt for women that I experienced at Sears. I can't count the number of times I still hear today from men, to whom I am earnestly explaining about a computer marketing or invention problem, "My, you are certainly as smart as you're good looking."

When I was eighteen and graduated from high school, I remember how proud my mother was. I also felt that education was the key to leading a successful life, to get out of the poverty rut our family had experienced.

In spite of my being an around-the-clock worker, I did well in high school; my grades were good. When I was in the eleventh grade, in the late seventies, I remember reading a book in the library that said the percentage of women making it in medical school was only two percent. So I thought, who am I to want to be a medical doctor, and recalled what that doctor's wife in La Jolla had told me years before.

When I graduated from high school, I was kind of lost about what I wanted to do next. I felt the need to achieve at a level far higher than what "girl jobs" offered me. It seemed so unfair to me. Looking back, I can now see that my feeling uncertain about what I should become prevented me from taking the easy way out. For the idea of marrying for money, marrying to help a husband's career and not my own, was repugnant to me. In that sense, I was already a product of the feminist revolution back in the late seventies, without even knowing it then!

Even though I didn't know what I was going to do with my life, I did know then that education and more education stood for the way out of the poverty I grew up in. I always remembered my father telling me, "You have to get an education to be successful in life." That's what I also heard from my grandmother even when I was a little kid.

My first college step was going to a junior college in San Diego while I was still working at Sears. I took childhood development; since I had raised two brothers, taking a course like that came natural to me. The teacher took all of us in that class to a hospital to observe children being born, and it was fascinating. The course dealt with observational and developmental methods as they applied to children and made me think about becoming a teacher. Everything I've done as far as my career goals has been based on observing children and creating programs and games to empower them.

One of my teachers said to me when I completed my two-year junior-college program, "Well, you've made it this far. So now you've got to go to San Diego State, since you have the talent to succeed." San Diego State is a four-year college, and with that encouragement I went there, majored in liberal arts studies, and obtained a teacher's credential. After I graduated, I taught elementary school children for three years. It was great! It gave me a lot of independence, and it was very creative. It gave me a lot of confidence, since it made me get used to talking in front of people (I had been quite shy).

I loved teaching so much I could have stayed there permanently. But due to lack of district funds, forty teachers were laid off. I had only three years of employment, so I was one of them since I had no tenure. When I left that job, the principal told me I would never come back to being a teacher again because all the good teachers like me were able to make more money in other professions. I didn't believe him because I loved teaching so much. But he was right. It's no different today— schools are still underfunded. In Silicon Valley one in three teachers leaves the area permanently within four years because of the high cost of living and the low pay. It's so hypocritical of all of the billionaires down there not caring at all about how children are being taught. They have a double standard because the people at the top send their own children to fancy private schools.

It was impossible then to get another job as a school teacher, so I decided to live on a sailboat to cut expenses while I was seeking employment. I didn't know how to sail, but it was something I had to learn if I wasn't to starve. Since I always had a talent for figuring out how to make things work, I started to learn to sail. I first tried to take navigation classes at the yacht club in the city, but they refused to enroll me because I was a female. So I took the classes in navigation at the junior college, which enabled me to use my small boat. That was in the early eighties when the talk about equal rights and opportunities for women appeared nightly on TV. It seemed so strange to me. On the one hand, I couldn't even take a navigation course because I was a woman; on the other hand, the media was saying we women were having it all!

It was at that time that I became a flight attendant. I worked for American Airlines from the age of twenty-four to twenty-eight and was stationed in San Francisco. I enjoyed the job, and it enabled me to see the rest of the world. I was assigned to first-class duty and met many business men who would tell me about their careers. One of them told me I could have been doing inside trading, because in one flight I would be talking to Microsoft or Intel people, and on a return flight I would be talking to Apple and other PC executives. I kept asking a lot of questions about their business because it seemed so interesting to me. And they were happy to tell me what they were doing. I think that it was at that time that I subconsciously felt that I would be happy working in the high-tech industry.

The salary at American Airlines was very low, so in order to survive I was living with five other girls. I became the head of flight attendant training and learned how to use computers there. When you're a flight attendant like I was, you have no personal life. People have the wrong idea that a flight attendant's life is a piece of cake. Yes, you work ten days a month, but you are always on call for the entire month. You have to carry a beeper with you at all times. There's no stability in your life and there's no time to build up a stable relationship. I could decorate a kitchen wall with all the business cards the men on my flights gave me. You're hit on all the time for dates. My flight attendant girlfriends and I went out with each other because we wouldn't date these business people. It wasn't comfortable to do so because you don't know who you're dealing with.

It's because I had no time to really know the person I eventually married, that our marriage ended quickly, in two years. I had a beautiful daughter, Jackie, out of that marriage and have been her sole support since I was awarded sole custody when she was two years old. She's now fourteen and has been a source of inspiration, and also a creator in her own right of many of the high-tech programs I developed. I should say *we* developed because she has been responsible for so many of the successful high-tech ideas for girls my Girl Tech Company created and marketed.

At that time I worked at a store in San Francisco, which was owned by a man who later founded Victoria's Secret. The store was called "My Child's Destiny." It was a retail store for families that was designed as a one-stop shopping center where you could buy flowers, shoes, clothes, books, toys, and computers. I managed the computer and technology department directed toward children's needs. Software developers would come into the store and ask for my advice. They wanted to know how the children played with the software games I was selling. They would show me the software games they were currently working on, and I would tell them right away what I thought about them and make suggestions as to what I thought would improve the games. They liked my ideas, and I then began to write for some magazines for parents about how technology and software programs could help their children's education.

While I was working at the store, I took courses at U.C. Berkeley and obtained a Computers in Education Certificate in 1986. I then got a job at Broderbund Software Company, which is a major entertainment and education software company located in San Rafael, California. I worked there from July 1988 to December 1991 as a product manager. I was responsible for product design and implementation of all graphic, sound, and programming elements, as well as overseeing quality control of the finished products.

There's this fantasy that everyone was making big bucks in the high-tech industry. Broderbund was a giant in the education and entertainment software at the time I was hired, so you might expect my salary would be in the high five figures or low six figures during the three years I worked there (I quit in December 1991). How wrong that media nonsense was! When Broderbund hired me, they paid me a teacher's salary—only $24,000 a year. I needed the job badly to support my young daughter and myself, so I was forced to take it, even though I could barely pay my rent. I was doing work equal to the men in the company who were making $70,000 and more a year. The boss of my unit was a woman. She was not listed as a vice president although she performed that function. She was a manager, and that was as high as a woman could get at that time. There were no women vice presidents at Broderbund, and she was pushed out of the company to make room for a male executive, who took her place even though she had been a very successful producer.

It was at Broderbund that I heard the luring pitch that's been repeated to prospective employees thousands of times in thousands of high-tech companies. I was told my job wasn't a nine-to-five, forty-hour-a-week job with overtime after forty hours. That was the old economy. In this new economy, I would have no fixed hours or fixed days of work. I was told, "You are responsible for whatever it takes to get the job done, so if you can do it in thirty or forty hours a week fine, but it's the job not the hours that count. Your wages might not be high, but you'll be getting bonuses and options so you'll grow as the company grows."

Well it was all BS! At the time I left Broderbund, I had been working at least ten hours a day, six days a week. (The company didn't even include a lunch hour as part of the working day!) The projects I was in charge of demanded endless hours of my personal life as well as office time. "Taking whatever time it takes to get the job done" was a cover story for my having to be on call twenty-four hours a day. My beeper and cell phone would be ready for a call from the company whether I was eating dinner at home or sleeping or taking my child to the zoo.

I dearly missed my young child. It felt like I was sacrificing my life for a company that didn't give a damn about me. When I asked for a raise to pay for the childcare my daughter needed, the company told me that childcare was my problem and responsibility. Broderbund is typical of the high-tech industry's attitude toward women and their childcare problems. A company that gives its women workers childcare assistance is the exception rather than the rule. The fact that Broderbund was paying me less than forty thousand a year after working there three years was all the company cared about. It was such hypocrisy, for Broderbund's reputation was based on how helpful and nurturing their products were for children. But on the other hand, they could care less about the children of the employees who were working for them!

My experience with Broderbund destroyed whatever naive opinions I held about high-tech companies truly caring about their employees' welfare. It also highlighted the gross inequality in the way women were treated in the industry. It's only been recently that women were given vice president jobs, and those are still few in number. We women are still in an employment ghetto, no matter how high our educational achievements have been. We're given all of the responsibility but no authority for the work we do. The so-called bonuses or options that we were to get turned out to be nothing but words. I left with nothing when I left Broderbund.

My Broderbund work experience is typical of the high-tech industry as it relates to women who want to succeed. Here it is eleven years after I left Broderbund, and the recent research findings show that women in the Internet economy earn 24 percent less than men and receive 50 percent less in bonuses than men.

After leaving Broderbund, I went on to a number of interviews with other companies, and the interviewers always asked me the same questions:

▼ Are you a team player?
▼ Will you do what it takes to get the job done?
▼ How many products can you handle successfully?

And yes, the final clincher: "You'll have full support in all the work you will be doing."

I never believed any of them. Translated into truthful language, these interviewers were telling me that I would work ten hours a day, maybe seven days a week, to get the job done; I would always be on call. A team player meant keep your team happy and productive, even if that means spending evenings with your programmers at a restaurant and feeding them drinks to make them work harder (even though you feel guilty about not giving your own child the attention and nurturing she needs). Yes, you're given the responsibility to complete your products. But you have twenty products assigned to you to develop, and if one of them fails, you're the bad guy. In other words, to put it bluntly, when the shit hits the fan, you're the one that has to take the hit, even though you may be totally faultless!

I always dreaded meetings—and they are endless in the high-tech field because the meetings were like a men's club since women were a very small minority. Unlike the men, women had to raise their hands to get a word in. And since women's voices are softer than men's, the men would feel what we were saying wasn't as important as what they were saying. The men at these meetings always felt it was their perfect right to interrupt what women were saying far more frequently than men. As for eye contact, we women always had a hard time getting the men to focus on what we were saying rather than on our legs and breasts.

There were other women in the industry who felt the same way I did, and a group of us decided that the only way to escape from being a second-class citizen in the high-tech industry was to establish a company of our own. We would make a real contribution to women of our own age and to all female children to wipe out the gender gap that was preventing females from achieving their maximum potential.

While we were banding together, I went back to school and got my doctorate in education at the University of San Francisco in 1995 in order to enable me to work skillfully and effectively to empower women and their children to achieve their maximum potential in life. My doctoral dissertation was titled "Perceived Elements of Gender Preference in Video Games Played by Secondary Grade Elementary School Children." Here are some of my basic findings:

"The attitude among second-grade boys seems to be that a fun technology game is for boys, yet girls might play it. This attitude appears to be shared by the technology game industry's executives. Most corporate technology game companies are dominated by males in the executive ranks. They are the decision makers. . . . Girls enjoy the product, yet are told it is not for them by the industry and media in the way they build and advertise the products specifically to boys. The peer group and family also reinforce the belief. Parents do not buy the products as often for their girls. Boys are more secure in the idea that the technology world is for them. . . .

"Society needs to recognize the mistake of limiting development and human resources potential when stereotyping technology use as male. Technology is a gateway to the future. Building equal opportunity for all children encourages respect and fosters communication skills creating a richer society. A change is needed in order to include girls in technology play. . . .

"The research supports the belief that change initiated by the industry and media level can have a positive impact on girls and boys in technology settings. . . .

"The high-tech games and toy producing industry should present products through the media in a way that will educate and lead to a new way of thinking about girls and technology. Adventure with respect to girls. Most ads and commercials reflect girls as passive participants in technology use. Boys are active and have power.

"Advertisers are contributing to children's self-perceptions. Avoid myths and stereotyped behavior and try to portray children in a more egalitarian approach to products and services. Avoid stereotypes and myths about females and males. The use of pink and pastel colors sometimes appeals to both girls and boys. Create products and marketing strategies that feature healthy female role models. Design products that are cooperative and open ended. Action oriented games are popular but children in this study actually preferred less danger and violence."

The research for my thesis reinforced my determination to accomplish the goals I outlined above. To that end, I organized a team of high-tech women who held my same convictions and established Girl Tech in 1995. Girl Tech is a multimedia company devoted to encouraging girls in technology use and is committed to building products and services just for them. The discrimination against girls is as evident in this market as the discrimination against their mothers who may be working in the high-tech industry. For example, boy-targeted products dominate the electronic toy industry. There are 18.6 million five- to fourteen-year-old girls in the United States, yet 85 percent of the revenues of this $10 billion market are derived from sales to or for boys. My research revealed that if technology products are created specifically with girls' preferences in mind, girls will buy and use them.

This sounds like conventional wisdom today, but back in 1995 it was considered revolutionary by the venture capitalists I tried to obtain seed money from to establish Girl Tech. While the venture capitalists on Sand Hill Road in Menlo Park were being written up in the media as if they were giving money away to everyone they met on the street, it simply wasn't true. And if you were a woman founding a company like I was, it was doubly hard to extract a nickel from them. I had to fund my company by credit cards and a $10,000 loan from a private backer. The venture capitalists told me to get lost. When I told them I was creating my company because I wanted to show girls that they are celebrated and valued in our society, that it's not about what you wear, but who you are, the venture capitalists looked at me as if I was some kind of a nut! Today, it's a different story. The success of Girl Tech—even in these downturn times—proves how wrong they were. You can create a product that sells and is also based on solid human values, instead of shock and sex titillations.

I recently completed my MBA at Dominican University. It's my eighth degree. I love learning and teaching, so I'll probably be taking courses when I'm ninety!

I'm forty-three now and never remarried. My daughter is now fourteen, and we are the best of pals. She's an excellent speaker and creator of toys in her own right and travels frequently with me when I tour the country. But I know she has her own life to live. I've never regretted taking full responsibility for her welfare. That combined with my work has been the total focus of my life. I've often been asked about my personal life when I lecture around the country. What personal life? That old saying that women have to work twice as hard to get to the same place as men is still all too true. I've learned that you have to be smarter, work harder, and put more hours in your work than men to succeed. And it's got to look like it comes easy to you, so you can't show how hard you've worked and all the time you put in to accomplish your projects. Otherwise, if you admit you work excessive hours, the men get the impression you're not using your time wisely. We women are plagued with this anxiety and the need to pretend. You not only have to work that hard, you also have to be available for work on weekends. Often, I was expected to fly on a Sunday to another part of the country to be ready for a meeting on Monday morning. So, you can say my personal life has been my business life. However, I feel I'm ready now in my early forties to start to nurture the other part of myself, the part that wants a loving companion, a caring and compassionate relationship with someone I can trust. It's true I have a hard time trusting men due to my business experience, but I know we are not alone on this planet. We women and men need each other. We need a partner; a desire to belong with somebody is not something to be ashamed of. There's a wonderful phrase that novelist Ford Maddox Ford wrote that I've always remembered. He said men and women need each other in a relationship that helps us renew each other's courage and enables us to cut asunder our difficulties. Mutual appreciation for the differences in our genders is a healthy way to live, not the kind of mean-spirited competition that characterizes Silicon Valley life. We're all vulnerable at some times and strong at others, so let's help each other rather than consider men and women as enemies in battle.

At this time in my life I want to be with my daughter more frequently; I want to hear all the gossip about her relationships with her friends. I want to be able to pick her up at school and spend time with her when we come home. That's so important to me now. I feel the need for a well-balanced, spiritual, healthy life. Before I ended up in business, I was a very spiritual, quiet, reflective person as well as being creative. But in business I felt

I was always on the front line of an eternal war. Right now I'm recovering from that war. I taught my daughter how to work and survive. But I also want her to have a good emotional life. We're making that journey together.

THE STRUGGLE TO BRIDGE THE GENDER GAP

As Janese Swanson so eloquently put it, the gender gap is an issue that's far more than being a money-gap. It is a gap in the mutual misunderstanding of men and women in the business world. It is a gap in the perception of women in business as being second-class citizens, with too many of the women themselves believing that's all they deserve to be.

Perhaps it should be called, above all else, "The Fairness Gap." For it has been the women's revolution that has focused on the fairness issue as being central to the need to end the battle between the sexes in business, as well as in all other arenas of life.

The culture of the high-tech industry is more deeply stained with gender discrimination than many other major industries because it was engineering and computer science based—two areas in which women were almost nonexistent at the creation of the industry. So it seemed like second nature for the scientists and engineers to carry their old male biases with them into the creation of the new cyber world. For example, the founder of *Wired* magazine, Louis Rosetto, a major influence in high-tech culture, stereotyped women as follows in a 1997 interview:

Women are biologically different from men. They have different hormones, they have different neurotransmitters, they have different body chemistry, they are created for different biological functions that lead them to be more inclined to do certain things. Men are content to sit in front of a screen for hours on end and do things with it, interact with inanimate objects, and men have been doing that, from computers back to cars, back to factory equipment, back to farm implement, back to whatever. Women are more social; they prefer to spend their time interacting with others than in solitary quests.

By Mr. Rosetto's standards, female computer scientists, design and quality engineers, software programmers, and CEOs, like Hewlett-Packard's Carly Florina, should be nonexistent in the industry today. (Women like Dr. Janese Swanson should be considered a freak of nature!)

But here it is five years after Mr. Rosetto's interview, and high-achieving women are now found in the very same jobs that Rosetto's definition of women's capabilities would have eliminated the possibility of their attainment.

It's true that stereotypes die hard and that women, now functioning at the higher levels in the high-tech industry, are still in the minority. But the emphasis on fairness for women in the workforce is having its positive effect.

THE NEW GENERATION
OF HIGH–TECH WOMEN

Pat interviewed Janese with me and has this to say about the fairness issue:

As a woman, I can immediately empathize with women like Janese and with the women I see in my counseling practice who are so much concerned about fairness—and the lack of it—where they work and in their personal relationships. The reality of our history as a nation has been from its very beginning the reality that women were seen as second-class citizens: women were denied the right to vote until 1920 and were defined as housewives, mothers, and caretakers, but never as career persons until the women's movement began changing that state of affairs. It is sobering to be reminded that the issue of women's beliefs in the essential unfairness of their position in society dates back before the Constitution was written. It was Abigail Adams, the wife of one U.S. president and mother of another, who proclaimed in a letter to her husband in 1776 that women resented "the tyranny of husbands," and that "if particular care and attention is not paid to the ladies we are determined to foment a rebellion. . . ." Well that rebellion took two hundred years to materialize in the form of the women's movement that originated in the 1970s.

Today fairness for women means equal employment opportunities in the working world outside the home; no sexual harassment on the job; both husband and wife are employed in all types of careers and jobs; marriage is an equal partnership, with the husband and wife mutually sharing household chores, and the husband is presumed to be an active partner in the birthing of and in caring for the children. All major decisions in the family are to be mutually discussed and arrived at in partnership.

It is my observation that we as women still have a long way to go to experience today's concept of fairness. More often than not, this principle is honored in the breach rather than in reality, as Janese's experiences in the high-tech industry vividly demonstrates.

However, I have great hope for the young generation of women now working in the high-tech industry. The educational impact of the women's movement's on today's working women cannot be underestimated. Although I've heard many younger working women claim they are not "feminists," they have incorporated in their own consciousness the fact that they are entitled to be treated as equals in the world of work and in society at large—and that they will fight to attain that fairness. Well, without the advent of the women's movement, that consciousness would never have arisen.

The women in the high-tech industry had been brainwashed to believe that they were working in an industry where all the rules about women being incapable of attaining high-level, equal-status, equal-income jobs with men were banished as being the product of the so-called old economy. But these women are very bright and are quick to detect that the reality of their working life is in many ways no different from their parents' generation and are taking very assertive action to correct the injustices they are experiencing in their workplaces. For validations of this fact, you need look no further than the long article that appeared in the *New York Times* of August 12, 2001, whose headline stated "New Economy, Same Harassment Problems."

The article reported that Littler Mendelson, an employment law firm based in San Francisco that has represented more than a thousand Internet-related businesses, estimates that two out of five managers of such companies will be targets of lawsuits in the coming years, twice the rate of the managers of bricks-and-mortar businesses. Complaints about hostile working environments, unequal treatment, wrongful termination, and sexual harassment are the issues women will be fighting to remedy. The article singled out Lori Park, a software engineer in her twenties who recently won a sexual harassment case against Juno Online Services. She alleged that Juno created a hostile work environment in which male supervisors considered women to be "sexual objects" and referred to many of them as "sluts" and "whores." She won her suit, persisting after experiencing attacks of depression as a result of the fierce resistance the company put up to her lawsuit. She now says, free of illusions about the new economy's attitude toward its employees, that she is proud of her effort to stand up to the company and said, "If I don't stand up for myself, who else will?"

This article ends with the following sentence: "Since her experience, Ms. Park said she has been informally counseling other young women who believed they had encountered harassment and discrimination at work on how to build a legal case against their employers."

To me, that's a most encouraging statement, since it is a sign that these high-tech women firmly believe that being treated fairly is essential to their working life—and that they will settle for nothing less. In my own counseling practice, I've dealt with many professional career women who have come to the same conclusion as Lori Park. They've learned that lesson from their own bitter experience. When I hear their stories, I'm reminded of the saying, "Whatever does not kill me, makes me stronger." These women are stronger.

THE DOT-COM
FIRESTORM
THE MONEY THAT WAS
ONLY CONGEALED SNOW

It was 1995, and the smell of money was in the air. There were millions—perhaps billions—to be made, and all you needed for the ticket that would win the lottery was an idea that seemed reasonable and the passionate dedication that would enable you to charm the venture capitalists to finance your potential fortune. Of course, the venture capitalists would demand 50 percent or more of the take and decision-making powers when they made the speculative money available to you. That would be part of the bargain, but it was agreeable. After all, there would be millions or billions to be made, so there would be more than enough to go around. No need to be *that* greedy!

Why was 1995 the year? It was crucial because it was the time of the creation: the creation of IPOs (initial public offerings) as stocks made available to the general public by venture capitalists in combination with founders of start-up high-tech companies at the right time and right place for huge stock-market gains.

Of course, IPOs were available prior to 1995, but on a very modest and limited basis. The successful high-tech companies in Silicon Valley, like Oracle and Hewlett-Packard, were "old-timers," being established companies that created fortunes for their owners in the years before 1995. They were substantial companies that created reputable, valuable services and equipment. On the other hand, most of the thousands of new high-tech start-ups that resulted in the creation of instant million-aires and billionaires were the stuff of dreams and media hype. They flourished in the short interval between 1995 and 2000, and with few exceptions all have disappeared. With few exceptions these "giant business moguls," as Mark Twain commented about similar entrepreneurs a century earlier, are "now fast rising from affluence to poverty."

What triggered five years of transcendent avarice was the initial public offering of a new company called Netscape, founded by Jim Clark who had been a computer sciences professor at Stanford as well as a technical innovator. Netscape was and is a significant high-tech company that created a storm of interest within the general public since its search engine permitted users to browse the Internet. It was Clark's idea to go public with a stock offering in 1995—and it was that IPO that made Clark an instant millionaire. On the first day of trading, the stock-market shares more than doubled, rising from $28 to $58.25. Three months later, that initial $28 share rose to $140.00. The American Dream seemed triumphant: Own the right stock and you, too, can become instantly rich and famous.

No need to work hard or be creative, just buy and own the stock that doubles or triples once you buy it.

Word spread like wildfire about the fabulous success of Netscape's stock offering, and soon young Wall Street investment salesmen—recently graduated MBAs and engineers from elite schools such as Harvard, Yale, and MIT—were salivating with dreams of instant wealth based on the Netscape IPO model. These dot-commers were bright and very young (mid-twenties would be the average age). They were men, not women, since high-tech was a macho industry (women engineers, technicians, and financial experts were notable by their absence). The women who were involved usually were ghettoized into human resources and marketing divisions—the rest of the operations were male-centered.

These men, who now saw the high-tech start-up company with its instant IPO as the road to instant easy success in life, were bright and obsessed with the idea of becoming millionaires who would retire by thirty. They saw, quite rightly, that they were at the right place at the right time: now was high-tech time; first is best, otherwise they would be toast. It was a correct observation. But it was also correct that even though this was indeed the right place and right time for an IPO (then take the IPO money and create a new start-up and a new IPO)—the possibilities seemed endless. It was also the time when the money they thought they would make became frozen snow that melted.

It was Dorothy Parker, that superb poet and screenwriter, who said, "Hollywood money isn't money. It's congealed snow, melts in your hand, and there you are." Pat and I have applied her "congealed snow" metaphor to start-up money because it's so apt. As Vern Anderson, CEO of Silicon Graphics told Michael Lewis, author of *The New New Thing*: Penguin 2000: "Silicon Valley has more in common with Hollywood than it does with Detroit. The venture capitalists are the studios. The managers are the directors. The ordinary engineers are the writers. And the entrepreneurs are the stars." What Mr. Anderson neglected to say was that most Hollywood productions are failures—and the same is true about the debris that so many ill-conceived start-up companies spawned. The dot-com firestorm in 2000 caused thousands of dot-com start-ups to vanish. That old phrase, "here today, gone tomorrow," was a reality when many of these start-ups which began only a year or two prior to March 2000 disappeared. Nearly a thousand dot-com start-ups have disappeared since March 2000.

What caused the disaster? Fear and greed is the answer. It's the same answer that created the monumental stock-market disaster of 1929. To understand the Silicon Valley firestorm, it's necessary to expose the cover story that Silicon Valley is driven by noble desires to better humankind. Jim Clark, the founder of Netscape, who triggered the start-up mania, has stated bluntly, "If I go and do the next thing and don't make $100 million, it'll be a failure. I'll be a failure." And an associate of Clark's says, "Jim Clark has a clarity of vision that is prompted by the purest form of greed. Nothing clouds it."

Strip away the romantic cover, and the real high-tech motivation for power and glory is the centuries' old motivation in high-tech clothing. When Microsoft survived after the antitrust government case against it, *Fortune* magazine (July 11, 2001) presented a close-up cover picture of Bill Gates with the title underneath announcing "The Beast Is Back." There once was a time in our country (think of Franklin Delano Roosevelt's time) when "beast" would be a pejorative designation of a cruel, voracious businessman. *Fortune* magazine uses it as a word of praise. And it is in this climate of opinion that the recent dot-com start-ups and failures proliferated.

There are two American traditions, clashing with and contradicting each other since 1776. There is the Declaration of Independence tradition of valuing "Life, Liberty, and the Pursuit of Happiness" as the foundation of the United States. And there is the countertradition of pursuing wealth at all costs as the only way to obtain happiness—and idolizing the business achievers and their fabulous wealth. The Jim Clarks of America echo the words of previous generations of business millionaires (the only difference today is that to be a famous business icon you have to be a "billionaire," since millionaire signifies small change in this twenty-first-century global economy). The difference may be in the bank account, but the value system is the same as when Lee Iacocca, a famous but now-forgotten business icon of the late seventies and early eighties, was widely quoted as saying, "That's the American way. If little kids don't aspire to make money like I did, what the hell good is this country?"

Mr. Iacocca, like today's high-tech CEOs, publicized that he was a self-made man, noted for turning the Chrysler Company into a successful business when it was tottering on the brink of bankruptcy. What he avoids telling the American people is that it was the federal government's guaranteeing huge loans to Chrysler that enabled the company to flourish. Without that government bailout, Chrysler would only be a memory of automobile history today.

We have already commented in a previous chapter that this "self-made man image" is a favorite pretend self-designation with which that the CEOs in the high-tech industry shower themselves. Yet they know very well that without the huge subsidies and contracts derived from the defense industry and other government agencies, plus costly government-sponsored research projects and tax breaks, plus immigration laws that allow them to hire cheap, foreign high-tech labor, the high-tech industry would only be a pebble in the international business pond.

THE SOIL IN WHICH THE START-UPS FLOURISHED AND PERISHED

There was a time when the label "start-up.com" was defined in the public's mind (courtesy of the media and the stock-market insiders' brainwashing techniques) as the road to instant wealth that everyone could obtain if they used their smarts. All you had to have was a saleable idea and some borrowed money to "start up" a company. What you produced and sold didn't matter, since once your company became a reality (in a garage, apartment, office) you could then create an initial public offering of your stock on the NASDAQ (the index of technology stocks) and then become an instant millionaire as the stock would double, triple, or quadruple overnight while you held most of the stock.

Or if you didn't want to create a start-up, you could just invest a few thousand dollars in the stock market, turn on the Internet or MSNBC each day, and see your high-tech stock always increase. It was evident that the stock would always rise because that's what it was doing every hour of every day back in the five-year period from 1995 to 2000. This would take you a little longer than creating a "start-up dot-com" to make your millions, but you could afford to wait, since your golden egg future was reassured.

We live in a time of what Silicon Valley denizens call "warp speed." In other words, it is time so speeded up and disorienting that industrial and economic developments that usually would take decades to mature, to rise and fall, now take a few years at the most. Consequently, what was just related about start-ups, which seemed like reality back in 1995, became a journey to nowhere by the end of 2000. And the average investor, who hoped to piggy-back on the stock market's escalating valuations of start-up stocks, found himself or herself wiped out—all in the space of one or two years.

Indeed, it was an historical achievement if a start-up lasted five years, for usually two years was the birth-to-death time span for the start-ups that left billions of dollars of investment funds in dot-com graveyards. This was warp speed in action. For the dictionary definition of "warp" is "to turn from a correct, healthy or true course; to pervert, to corrupt." The consequences of actions driven by fear and greed are found in dot-com graveyards, and the vanished bank accounts of the average stock-market trader who thought productive work was a frivolous way to make money.

Many boring books written in tedious economic jargon have been written about this dot-com and stock-market debacle. Our is not one of them, for it is the fundamental untamed drives of greed and fear and their effect on the quality of life that is its central focus. For it is those drives that led to the misreading or ignoring the signs that there was danger ahead, rather than success at the end of the line for the start-up dot-coms.

As witnesses of the start-up debacle, the general public is fortunate to be able to view a recent documentary film dealing with Silicon Valley failures. It is called *Startup.com*, a factual film that follows two young (mid-twenties) men who pursue their dream of becoming instantly wealthy by establishing a start-up company, and how that dream became a nightmare, within the warp-speed time period of May 1999 to December 2000. These two entrepreneurs typify the start-up mentality and age level of the men (almost never women) who drove their business firms from launch time to disaster. Although they lived on the East Coast, they are no different from the hundreds of entrepreneur hopefuls Pat and I have interviewed here in Silicon Valley. Their mentality and their actions are the same. First and foremost is the dream of instant wealth. They are caught up in the speculative frenzy that began in 1995. As eminent economist John Kenneth Galbraith remarked in a TV interview: "Speculation on a large scale requires a pervasive sense of confidence and optimism and conviction that ordinary people are meant to be rich."

The cofounders of a start-up called GovWorks.com were Kaleil Isaza Tuzman and Tom Herman, and at the time they

founded the company in May 1999 they were inseparable
friends since high school. Kaleil was the CEO, and Tom the
technical manager. Of course, they didn't consider themselves
"ordinary" people. They came from upper-middle-class backgrounds, and Kaleil had just quit his high paying job at the prestigious Wall Street firm of Goldman Sachs. But he and Tom felt
"ordinary" because they hadn't yet become millionaires at the
age of twenty-eight. They had an idea they thought was the
key to fame and fortune: they would establish a Web site that
made it easier to pay parking tickets, pay taxes, and buy licenses
through their GovWorks Web site. They kept convincing themselves that they were right and speculated how wonderful it
would be for people to use their Web site at three in the
morning to buy a fishing license from home or attend a town
meeting from a Web site while in their underwear.

A careful, realistic evaluation of their idea would have
revealed it was a wet dream rather than a sensible, profit-
making, long-term business project. But it was 1998 when the
two friends created this proposal. It was within the five-year
period (1995–2000) when literally the wildest business dreams
could get financing if they were presented with the passion and
sincerity that a first-class used-car salesman possesses. Certainly
Tuzman met these criteria.

The film exposes the charismatic charm of Tuzman as he
tirelessly travels throughout the United States trying to obtain
venture capitalist financing for GovWorks. He is the pitchman
and Tom Herman his technical backstop in these business
pitches. The presentations are so absurd it's hard to believe they
were allowed the time, effort, and consideration of cool, poker-

faced venture capitalists who were twice their age. These are the same capitalists who were tight-fisted with a dime before the dot-com gold rush began in the mid-1990s.

Since this was the period of fear and greed, Tuzman and Herman were in the right place at the right time to get the financing they needed. For the venture capitalists, indeed, did give them sixty million dollars to finance their company. The money was advanced out of fear and greed: fear on the part of the venture capitalists that if they didn't advance the money, they would miss being in on a giant opportunity in a new revolutionary industry, where being first is best. Greed was the engine that fear fueled. Investing in a component of the new, fabulous high-tech invention known as the Internet—and reaping an instant fortune as a consequence—was the vision that blinded their eyes.

In the warp speed time of less than two years the tangible sixty-million-dollar investment turned into congealed snow. As the film documents their business descent into oblivion, the two entrepreneurs are shown in psychologically naked disarray, in contrast to their euphoric beginning. When their start-up began, there is a scene in a cab where both of them look at each other in amazement, with supernal delight in their faces. They say to each other: "We're going to become billionaires!" That is the high point of their ideals. It is evident then that creating a company as a public service had been the farthest thing from their minds. The money and the power it gives was their driving force. (The curtain of public benefit is dropped in another scene in the film where a board of directors member speaks on how valuable a service the company is giving, and then says at the end, "And of course, make a lot of money!")

THE DOT-COM **FIRESTORM**

Later they find that their business service is ignored by Internet users and GovWorks.com loses millions of dollars each month. Terror grips them. Tuzman puts his hands together in prayer, and it's evident that he is not praying for peace in the world but for the survival of his business income. Financial survival is now the obsession of Tuzman and Herman, where once it was being billionaires. They cannot get any new financing since the company has become a black hole instead of a profit-making machine. (GovWorks had never made a profit; it only generated the illusion that it would eventually make a profit. This is typical of all the start-ups that self-destructed at the start of the new millennium.)

Their obsession to become billionaires had defined their identity as human beings. And when they saw that dream of who they were to become vanish, the fallout from their disillusion began. It took the form of a bitter breakup between two young men who were practically blood brothers since their high school days. To preserve his CEO status, Tuzman fired Herman, and Herman left in bitter, stunned disbelief. Tuzman's girlfriend broke up with him, for he was devoting all his time to business, and she felt irrelevant.

The end of this documentary film leaves the two entrepreneurs as shells of human beings. They had invested their entire sense of themselves as men who had the potential of being admired for their wealth and power, but all they were left with was congealed snow in their hands.

I have related this story of GovWorks' rise and fall because it is typical of the thousands of Silicon Valley start-ups that originated between 1995 and 2000. Pat and I have seen this

113·

movie often, with many young entrepreneurs of similar age and background and start-up experience as Tuzman and Herman. Their responses are always the same: It's the response of "I've been there and done that, too." These entrepreneurs wince when they see the two, in effect, begging for a handout from venture capitalists. They nod sadly in recognition of the disarray in their own lives, of the kind that left Tuzman and Herman with no time for nourishing family, friends, and a love relationship. They wince when they see the scene in the film when the GovWorks officials lead their workforce in a cheer designed to enhance employee motivation and production efforts. The officials shout:

"What are we going to do? Rock 'em."
"When are we going to do it? Every day."
"How are we going to do it? Every way!"

As one of the entrepreneurs who saw the film told us, "Yeah, we used to have a similar cheer at our place. It was like getting false courage from a line of cocaine."

To the young entrepreneurs we talked to, the most sobering moment in the film was the exasperated comment of Tuzman's neglected girlfriend, who sees through the facade of her boyfriend's CEO image. She says: "They look like grownup gentlemen, in their ties and their suits. But they're not. They're just kids."

As for the bitter breakup between the two childhood friends who founded the company, Jonathan Weber, a reporter for the authoritative high-tech magazine *The Standard* reflects:

"*Startup.com* zeros in on the relationship between the two main founders, who are childhood friends. The lesson is that in the end, a company is about business, and business is cruel and unforgiving and at some level incompatible with true friendship. . . ."

It is ironic that Jonathan Weber wrote these comments in the May 7, 2001, issue of *The Standard*, because only three months after his article appeared, this dot-com publication went bankrupt—another example of the "cruel and unforgiving" nature of business!

A SAMPLER CHRONICLE OF START-UP DOT-COM DISASTERS FORETOLD

If you log on to a search engine such as Google or Yahoo and type in the phrase "dotcom graveyard," you will find a plethora of start-up failures that are both similar and different from GovWorks.

They are all similar to GovWorks in that they also failed in warp time—two or three years from their inception for almost all of them. They are similar in that they, too, were mostly companies that existed solely on the Internet without any brick-and-mortar physical structure. In other words, these start-ups were ideas, based on imagined services that the entrepreneurs thought millions of customers wanted. Millions of dollars of venture capital was poured into these black holes of business. These companies, like GovWorks, were ill-conceived, badly executed businesses whose belief in consumer acceptance was wildly misplaced. The products they sold over the Internet were either irrelevant or could more easily be purchased in regular brick-and-mortar stores. Many of them were simply "concept companies." For example, Pets.com lasted two years on the concept that selling pet supplies online was the wave of the future. But consumers thought otherwise. Pet lovers preferred buying their cats or dogs their food and trinkets in regular stores. The result was that Pets.com lost *$61 million*, with revenues of only $5.8 million. Richard Villars, a dot-com analyst for International Data Corporation, remarked about Pets.com and hundreds of similar concepts, that "It's one thing to have a dumb idea. It's another to have a dumb idea in front of a national audience."

Here were some of the ideas that were taken seriously by sober financiers: selling cereal (Flake.com); high-fashion retailing (Boo.com, which lost *$250 million* in venture capital); selling toys more easily than purchased in a neighborhood store or mall (e-toys, Inc.); selling beauty products (Eve.com); or simply selling nothing! (According to a Securities and Exchange Commission report, NetJ.com was listed on the stock market even though it admitted, "The company is not currently engaged in any substantial business activity and has no plans to engage in any such activity in the foreseeable future.")

The start-up debacle cannot be attributed solely to the youthful indiscretions of budding entrepreneurs like Kaleil Isaza Tuzman and Tom Herman. For none of these start-ups could have appeared without the backing of middle-aged bankers and other men who advanced the financing.

Many of the start-ups that failed originated in the minds of middle-agers rather than kids who looked like recent high school graduates. For example, Pop.com was announced with great fanfare in the media centers of the world in 1999. It was a Web site founded by Steven Spielberg and his associates at DreamWorks, as well as movie director Ron Howard, the founder of Image Entertainment. Pop.com was supposed to create short-feature entertainment and would get its profits from online advertising. But advertisers stayed away from Pop.com in droves because it didn't pull in customers, and the founders decided to end the project. Pop.com was as poorly thought out as the many hundreds of other dot-com failures before and after it. Yet some of the smartest, wealthiest business moguls funded it. All of the originators of this failure were middle-aged.

Similarly, Webvan.com, one of the greatest start-up failures in history, was a product of middle-age hunger for profits. It was the idea of Louis Borders, creator of the Borders book chain, to build an Internet national online grocery delivery store called Webvan. Because Borders was successful in creating his brick-and-mortar bookstores, there was a tendency to believe he could be successful in any kind of business. Apparently, this may have happened to Louis Borders, even though an Internet online Web site service was vastly different in its problems and possibilities than selling books in physical bookstores. Since Borders had the cachet of being an entrepreneurial success, there was little difficulty in raising huge sums of money—$800 million to be exact—from seasoned investors such as Softbank Venture Capital ($157.9 million), Goldman Sachs ($100 million), Amazon.com ($62.5 million), among many other middle-aged stock-market players. There were some reservations among other venture capitalists that Webvan was a flawed, ill-conceived idea, but fear and greed triumphed over reason: fear that this grocery-delivery service was the first of its kind in the country, so that there was fear of losing millions of dollars if you weren't first; second there was fear that they would be too late to reap windfall profits. Greed blinded the investors to the very real and obvious flaws in the Webvan idea. The decline of its stock price mirrors the disastrous descent of Webvan into oblivion.

Webvan's initial stock offering rose from $15 to $25 per share on its first day in November 1999. But by July 6, 2001, its stock was worth only *six cents* per share. In the time of less than two years, $800 million became congealed snow.

It's well to question the expertise of Silicon Valley's CEO billionaires. For it is open for exploration as to whether or not the fortunes they made were more a matter of luck than ability. Of course, they had ability, but it is an open question as to whether or not being in the right place at the right time and their insight to take advantage of that fact made them the billionaires they became. But can that lucky opportunity be more than a once-in-a-lifetime opportunity? The investments of these business moguls in so many failed start-up companies like Webvan make this a moot issue.

The virus of hubris—pride, presumption, and arrogance— inheres in individuals who seem to equate the size of their bank accounts with their view of themselves as possessors of superhuman qualities, only to discover that their ability to transform the universe was severely limited. Ironically, up until the recent Silicon Valley economic disaster, "arrogance" was trumpeted as a badge of honor, as a sign of achievement in the community. "Silicon Valley arrogance" was the term Pat and I heard countless times during our research, and we always heard it said with overtones of entitlement and self-praise by the possessors of fortunate business positions.

This strength of hubris is different from the greed of the twenty-somethings with their start-up visions of instant fortunes that became only smoke and mirrors. All the twenty-somethings wanted was to be billionaires. But what of the Silicon Valley wealthy business elite, the established middle-age seniors who had "made it" before the 2000 debacle? Their greed took the form of believing they truly could transform the world, not with money (they already had *that!*) but with the force of their ideas. Perhaps the best cautionary tale as to the limits of their ability to do so is evidenced in the rise and fall of Palo Alto–based Interval Research Co. Under the title "The Think Tank That Tanked" the *Silicon Valley* magazine's September 3, 2000, issue told the story of hubris that typified the cultural environment of Silicon Valley before the crash:

> It was supposed to be Shangri-La, Silicon Valley style.
>
> Imagine: The world's second richest man (Paul Allen, Co-founder of Microsoft), pledges $100 million and a 10-year commitment to build what he hopes will be the most ambitious research venture in valley history. He hires some of the greatest minds on Earth and charges them with inventing the future.
>
> Shrouded in secrecy, cloistered away from competitors, they will innovate Paul Allen's vision of the next great revo-

lution after the personal computer, his futuristic Wired World in which consumers will receive information anytime, any place on demand.

The billionaire co-founder of Microsoft wants his elite research troop—unlike the corporate labs of IBM, Xerox and Microsoft—to be free from commercial pressures so they can pursue that valley rarity, pure research. Only after four or five years will they have to start spinning ideas into profitable start-ups.

This was 1992; the Internet and the World Wide Web as mass communication tools were in their infancy. Allen felt the personal computing revolution was stagnating. So his idealistic Palo Alto–based lab, Interval Research Corp. would leverage this "interval" between PCs and the next great wave, looking over the horizon and designing technology that would have an impact in five to 10 years and drive the computer industry for as long as 20 years.

That was the dream.

Eight years and a staggering $250 million later, Shangri-La is a Silicon Valley ghost town. On April 21, 2000, Paul Allen prematurely—and with little fanfare—killed his grand experiment. Once touted as a bold incubator of innovation, Interval is now reduced to a handful of researchers working on commercial applications for Allen's portfolio of cable and broadband ventures.

The story of Interval's demise is a cautionary tale not just for researchers, but for all entrepreneurs and Silicon Valley dreamers, revealing how the best brains, wads of money and abundant creative freedom can implode, leaving behind a stack of unexploited patents, accusations of betrayal, damaged careers and failed start-ups. . . .

Interval's fall from grace is a dramatic example of Silicon Valley arrogance. That arrogance has been accurately identified by John Seely Brown, chief scientist of Xerox Corporation. When he was asked if there was a Silicon Valley information age ideology, he replied:

"There is absolutely an ideology of the information age, basically faster is better, more is better, the whole sense that high technologists are kind of the high priests of society and we kind of know what is right. I think that there is a certain arrogance that builds up around Silicon Valley, that is that we understand how to make economies work; we understand what's right for people in terms of the technology of today and tomorrow; we understand what their future is going to be."

There was no indication in this eminent scientist's statement that the world economic destabilization we are experiencing today would occur.

The Silicon Valley elite technologists would do well to remember the wise words of Johann Goethe, the great writer-dramatist:

"Truth is contrary to our nature, for truth demands of us that we should recognize ourselves as limited!"

Goethe wrote these words over two hundred years ago, yet their wisdom resonates for today —and for tomorrow, too.

chapter
SIX

THE QUALITY OF
PERSONAL LIFE IN
SILICON VALLEY
LOOKING FOR HAPPINESS
IN ALL THE WRONG PLACES

"We work so consistently to disguise ourselves to others that we end by being disguised to ourselves."

—François, Duc de la Rochenfoucauld

The subtitle of our book is "The High Cost of the High-Tech Dream." Pat and I have been well aware of this high cost for many years *prior* to the recent disappearance of companies and the severe loss of jobs in Silicon Valley. The high cost to people's lives (the CEOs, the managers, the engineers, the assemblers, the service workers) has been enormous *during* the time of full employment, high incomes, and phenomenal profits evidenced between 1995 and 2000.

We are psychological counselors and at our Creative Divorce, Love and Marriage Counseling Center we have specialized in helping men and women and their families improve the quality of their lives. And since I am also an economist, cognizant of the significant social and economic transformations created by the high-tech industry, many men and women working in Silicon Valley have utilized our services in trying to cope with the disarray of living in their community.

During the period of high prosperity and full employment, Pat and I were faced with an increasing number of mid-level and upper-level management employees from Silicon Valley requiring our services. Somehow things weren't working out for them as they had predicted (and were told by all the publicity fanfare about Silicon Valley).

Why did they come to see us? Shouldn't they have all been happy according to their stock-market investments, their stock options, their year-end bonuses, and their ever-increasing numbers? Pat and I had been to parties during that period where many of these dot-commers were present. We were always amused when a self-important young man or woman would introduce him- or herself by handing us a card that would list himself or herself as the CEO of this or that start-up company. The title CEO, they believed, proved they were very important people. And there was the smiling young twenty-something-year-old woman, who introduced herself to us by saying, "My name's Joan, and I'm a millionaire!"

That was the temper of those times in Silicon Valley for the select, well-educated, predominantly elite school of graduates who had been nurtured in very comfortable middle- or upper-class homes. Yet many of these same and similar people sought us out for counseling since they felt their lives were spinning out of control. They felt there was a disconnect, an out-of-balance quality to the way they were living their lives. It wasn't fulfilling is what they would be telling us about the nature of their everyday lives. They would tell us this with a mixture of genuine confusion tarnished with apologetic guilt. They were succeeding yet not succeeding—a case of "be careful about what you are asking for, you're liable to get it." In one form or another, they were feeling depressed: while the stock market soared, their souls were plummeting. They were unhappy, but the media was telling them that the size of their bank accounts and the prestigious jobs they held qualified them to be exceptionally happy. They were feeling guilty because they "should" be happy by the standard our society dictates for happiness. And because they weren't, they felt guilty, since they played by society's rules and yet did not receive the emotional benefits they thought they were entitled to. They had conformed to society's criteria of happiness (that is, wealth and a good job as the tickets of entrance to an exciting life filled with love, emotional satisfaction, and a future of ever-expanding possibilities for triumphing over new challenges life would present in the subsequent years).

The men and women we saw who expressed to us their depressive feelings in the face of material success thought they were exceptions to the rule and felt something was wrong with them for feeling the way they did. Their request to us was simple: Could we help them feel happy, like they did when they first held jobs in Silicon Valley? Could we make their discomforting feelings disappear? Maybe, they told us, all they needed was a pill prescription—take the right pill and all their troubles would vanish.

They were looking for answers in all the wrong places. For they weren't abnormal people (as they themselves thought they

were) but were all too normal! For unhappiness was—and is—
a national problem, not a sign of individual failure. This fact of
American life received unequivocal confirmation in the path-
breaking Surgeon General's Report on Mental Health that the
government released in the heart of the economic boom, in
1999. It reported that "about one in five Americans experiences
a mental disorder in the course of a year. Approximately 15
percent of all adults who have a mental disorder in one year
also experience a co-occurring substance (alcohol or another
drug) use disorder, which complicates treatment."

In this context, the report revealed that after heart disease,
major depression ranks as the most severe disabling condition
in the United States. And the rich, no less than the poor, are
afflicted with this condition. The findings of Dr. Jeffrey Lyon
Spelles and Dr. Tanya Koskosz, who have studied depression in
corporate settings, estimate that as many as 10 percent of senior
executives have at least some symptoms of manic depression,
yet 9 out of 10 of their cases are going undiagnosed and un-
treated, according to a *New York Times* report on July 21, 2001.

There is irony in the reality that during the lush economic
years of 1995–2000 serious emotional distress was present
among those people you would expect should be the happiest
people on Earth. For it was the time when the media shouted
that you couldn't walk through a street in Silicon Valley's fabled
cities, such as Palo Alto, San Jose, Mountain View, Sunnyvale, and
Redwood City, without bumping into a new instant millionaire. It was the time when the dreams they lived by, which were
their stock ownership and stock options, made them wealthy
beyond their wildest dreams. So before their stocks and options

turned into congealed snow in the years after 2000, many of them actually lived the lives of the millionaires they thought they were: there were 20,000-square-foot monster houses they purchased, yachts, jet planes, million-dollar parties in Las Vegas, and sex with the highest quality prostitutes if that's what they desired ($10,000 a night of "fun" was not exceptional). Wasn't this the American Dream made a reality? For this American Dream was the imprint our media tried to impose on the psyche of everyone who regularly read a magazine, newspaper, or looked at TV programs. It was the dream of more is better, more money that is, which would buy the total freedom to indulge in the wildest fantasies one could imagine for oneself. It was the dream of buying one's way into total happiness.

Certainly, it looked reasonable: See the TV pictures of the superexpensive house in the lush exclusive neighborhood, see the multimillion-dollar yachts competing with each other in races only the superrich were entitled to enter, see the private jets, see the self-indulgent parties with the beautiful women hanging onto the arms of average-looking men, whose money was beautiful. Every major magazine from *People* to *Fortune* was littered with praise and pictures of these fabulous events, whose ticket of entrance was the amount of money one had at one's disposal.

However, what the media neglected to mention was that this elite segment of Silicon Valley was only a tiny fraction of the American population—less than 1 percent, to be specific!

But the image of towering wealth and self-indulgent men (women were the exceptions) wallowing in a world the rest of the country saw as an impossible dream for itself proved irresistible for the media to ignore. Exploiting these images created higher TV ratings, greater magazine and newspaper sales, and Internet business which enhanced advertising revenues.

And yet Silicon Valley proved to be a sea of excitement that generated mass unhappiness even in those five years of super abundance between 1995 and 2000. First of all, it was an unhappy time for the overwhelming majority of service workers, assemblers, and immigrant laborers who constituted the majority of the Silicon Valley workforce. The working homeless in Silicon Valley's Santa Clara County amounted to fifteen thousand in 1995, yet in the time of its highest prosperity in 1999, twenty thousand were homeless. It was a new kind of homelessness—a homelessness of hardworking employed men and women who couldn't afford to rent a tiny apartment. And in places like East Palo Alto, next door to elite high-tech Palo Alto employees, it was quite typical—and still is—to find three families living in apartments designed for one-family living.

Consequently, for the majority of the people working in Silicon Valley, the area has been a place of *diminished* hope, where the best that could be expected even in flourishing times was a minimum-level existence. People who worked as they did for wages of $10.00 an hour or less, sixty or seventy hours a week (and their spouses doing the same if they were married), hardly had the time in those affluent years to think of anything else but surviving until the next day. Under such circumstances, when you would ask them how they enjoyed living in such a wealthy area as Silicon Valley, you would get a blank stare or quizzical look, as if to say, don't you know that just making ends meet is our only concern!

But what about the "elite" engineers making $75,000 to

$150,000 a year, with stock options that presumably made them millionaires on paper in that time? Why were so many of them disoriented, uncomfortable, anxiety-ridden, and sad? Pat and I soon found out when they came to see us for psychotherapeutic help:

▼ They were working eighteen hours a day, often seven days a week, and the stress was beginning to tell on their health.

▼ They were always on call even in their off hours because of emergency situations.

▼ Their entire lives were invested in their careers: to be a "geek" was a badge of honor to them rather than a term of opprobrium.

▼ The companies they worked for had become their only communities. For there was no need to meet new and different people, commit oneself to new interests, or develop new relationships when the company that employed you provided a ready-made community for you. The company became a self-centered world in which you, as an elite engineer, could indulge. Take your pick: the company gave you a gym, a baseball or basketball team, a swimming pool, a hot tub, a cut-rate shopping store, a Friday-night beer bust, and social gatherings in which you could continue to talk about the corporate projects you were working on all day (and night also).

They told us about what we have termed, the "Been-There-Done-That-Now-What?" Syndrome. They had captured the

American Dream, hadn't they? It was all about money and the entitlements money could buy, wasn't it? It was all about status, about being valued for the "cool" career you established by yourself: they were the successes, all others failures, weren't they? So they should be happy, yet why were they so depressed? The symptoms of depression were defined in what they told us: sleepless nights, anxiety attacks, boredom, work projects that used to be exciting now seemed dull, anger over the pressure to produce more and more in shorter periods of time—and being unappreciated for doing so. And the need to self-medicate their anxiety, stress, and dead-end feelings with alcohol and/or amphetamines and tranquilizers was escalating to the level of addiction. If excitement couldn't be generated by the lifestyle they had opted for, then cocaine and crack were becoming endemic in their work communities. Drugs combined with beer-and-pizza dinners made their physical condition deteriorate.

There were two underlying causes of this condition they found themselves in. The first was the self-deception of believing that super material wealth was the definition of happiness. They had confused the American Dream of life, liberty, and the pursuit of happiness with the pursuit of money for its own sake. If you believe your bank account is your *sole* identity, then you will be looking for happiness in all the wrong places (for example, in buying a mansion, associating with celebrities, tripping around the world in an orgy of self-gratification, or the money-bought sex that created the fleeting excitement of a hit of cocaine).

The companies they worked for created wombs in which all their creature-comfort needs could be satisfied. It seemed great at the time they started their employment, but three or four years later (the time they began to seek our counseling services) it began to seem strangulating. In our psychotherapeutic counseling sessions, these elite engineers, graphic designers, information technology managers, and human resources directors began to understand that a career was just a career, and a company was only a company, not a lifetime culture. For it is the very nature of a company to be solely concerned about its profits and stock-market price. A company's loyalty is to its own bottom line, not the welfare of its workforce. If the welfare of its workforce enhances the bottom line, fine. If not, increased productivity demands and downsizing will substitute for workforce creature comforts. This is the reality of the business world, not to be condemned, but to be understood instead. It is one thing for employees to cope skillfully with this reality, rather than to deceive themselves into thinking that living a life means living in and for the companies they work for and nothing else.

There were warning signals in the affluent 1995–2000 that the quality of life demanded more than a dedication to one's career and company. Career and company employment indeed are important, but not the total definition of a human being. To believe otherwise was the dilemma the men and women who sought our counseling found themselves in. There was a national program that appeared on PBS in 1998 that drew attention to this dilemma, but was overlooked since the national obsession with rising stock-market prices was reaching its height then. The program was called "Escape from Affluenza." It defined "affluenza" as follows:

"Millions of Americans are falling victims to a debilitating disease called Affluenza. Its symptoms include swollen expectations, shopping fever, chronic stress, frustrated families, festering social scars and a painful addiction to the relentless pursuit of more—and if it is not treated it leads to perpetual discontent."

Silicon Valley represented this "relentless pursuit of more" to an unparalleled degree; more work and less time for anything else was its way of life. No time for raising children, nurturing a live relationship, or enjoying family life. Getting ahead in business really meant giving up the rest of your life.

The men and women we counseled were really saying to themselves, "I'm much more than my work!" It was a cry for help to find out what that "much more" really was and how they could actualize their needs beyond just working, no matter how tantalizing the money goal was.

They learned that in order to get "much more" out of their lives, they had to settle for much less! Much less, that is, of their obsession with work and income as the total definer of their identities.

When we asked them what they wanted in their lives *now*, not when they were first hired, they told us almost in unison, a happy family life, a lasting love relationship, and—if they were single—children and a spouse with whom to share a lifetime of growth together.

It was like they were reinventing the wheel! For every public opinion poll in recent years ranging from Gallup and Harris polls to *USA Today* and the *New York Times* has revealed that the overwhelming majority of both men and women rank their relationships as the number one priority in their lives.

Money and sex rank at the bottom level of every one of the polls taken. Good mental and physical health arise out of good relationships. Connectedness to others, service to one's community, and the flexibility to engage in new challenges in life are the sources for self-renewal. It's not simply a change in jobs that's the cure, it's a change in one's value system. When your calendar manages you instead of you managing your calendar, then a red light signal is flashing! It's when you have to chose between being a team player and being a family man that another red light flashes. It's when you're on call twenty-four hours a day and get called back on the job when you're supposed to take your spouse to a Sunday picnic that you're being warned something is off track. It's when you're missing out on the times when your child takes a first step or starts to talk that questions of what's valuable arises. If the price is too high, it's time to move on. If the price can be balanced with a life beyond work, that's fine. The choice is up to each individual who feels he or she is on a collision course with his or her own life.

A career and a well-paying job are no substitutes for bonding with a loved one and creating a shared future with him or her. We found ample proof of this fact among the men and women who presumably were "making it" and yet came to us for counseling because they felt isolated and lonely and felt guilty because they "had it all" and believed they therefore had no right to feel unhappy. You may recall the tremendous media publicity Silicon Valley received in those plush times from 1995 to 2000 that pictured high-flying, high-tech employees and CEOs taking their dogs to work with them. What the public was not shown was the unhappiness of those dog lovers. For the dogs were substitutes for the love they wanted, needed, yearned for, yet feared at the very same time. We would give these men and women a copy of an article we had written and wait for their response. The article read as follows:

When Your Life Goes to the Dogs

It's *real* dogs we're talking about! This happens when you feel your life is empty, when a satisfactory long-term relationship never existed in your life and yet the need for such a relationship keeps persisting. So you seek out a substitute lover—and what better lover exists than a dog.

President Truman was once asked how to find a true friend that would never reject you in Washington (read "Silicon Valley" today). His answer was, "Buy a dog!"

Just think about what a dog can give you: a dog can give you unconditional love, lick your hand and cuddle with you when you feel threatened by business competitors or a stock-market downturn or a layoff possibility.

Your dog doesn't care if you have money in the bank or are overwhelmed by debt. He doesn't even care if you have millions of dollars to spare! Under either circumstance, he or she will look you in the eyes with adoration, reassuring you you're a good person just because you're you. . . .

And who is the first one to jump on your bed and snuggle with you in those three-o'clock-in-the-morning hours when a chill in the soul might suddenly occur? Your dog.

There is only one fault in this "ideal" situation: your dog can only love you as a dog, not as a human being. You will have to look elsewhere for that. When your dog sits beside you as you watch your TV shows, the house or apartment you live in will consist of only one *human* being. Two human beings living together and enhanced in their happiness by the dog beside them—now that's a concept!

Some of our clients would smile and nod in recognition, while others would remain silent and rather sad looking. In either case, they would acknowledge that, yes, that's what they were doing, using their dogs as a substitute for a relationship rather than an enhancement of one. In the very act of their reading the article, they gave themselves permission to acknowledge their own vulnerability: it was not a sign of personal failure or weakness, but an honorable expression of being human, not a business machine. In this world, people need each other, not as a threat but as a promise of happiness. The great poet, W. H. Auden, summed up the human condition well. He wrote: "We must love one another or die."

WHO AM I? THE LOSS OF IDENTITY

Since what we have described above was the state of personal well-being in Silicon Valley *before* the severe recession that began in the middle of 2000, what has happened since then? In a previous chapter, we have described the shock and surprise, the disorientation and disarray of the presumed elite careerists in our society, the high-tech skilled workers. Suddenly their sense of entitlement vanished when their jobs began to vanish and no new jobs materialized for them. The vagaries of the job market could easily be contended with in flush times, since

another job was right around the corner if a project terminated. But now it was six months, then nine, then longer, and
no new jobs were in sight. Their stock options proved to be a
source of nothingness, rather than a cushion of comfort in a
bad time. Credit cards were maxed, expensive cars sold, houses
put up for sale (at a loss), and suddenly Silicon Valley became a
place that invited you to disappear from its environs. The thousands of men and women who came to Silicon Valley in gold
rush times now discovered that they came to a desert instead.

The high-tech men and women who came to Silicon Valley
in their mid-twenties, who only possessed an MBA or a Ph.D.
and no business experience, were the people who experienced
the most severe identity crises. Older people, who had ten or
twenty years' experience with the Valley's employment and unemployment conditions, could experience the new downturn
with a "been-there-before" attitude, although they became
worried (and are worried today) over the long-lasting downturn of the industry. Maybe this time they, too, wouldn't be able
to weather the extended storm of unemployment that current
economic conditions foretold.

The twenty-somethings, not having developed a long-term
sense of self that enabled the older men and women to prevail
over unexpected economic tribulations, were vulnerable to psychological onslaughts on their sense of self. Most of these
twenty-somethings came from comfortable middle-class
homes; they went to first-rate colleges and universities and took
proud comfort in their MBA and Ph.D. titles. They had graduated in complicated technological arenas and felt smug about
that fact. Indeed they were unique, but they were unique in

only one limited area of experience: they had a skillful relationship with *things*; they delighted in solving technical programs, repairing, building, and creating new tools to make things work in the burgeoning field of information technology; they loved working on "technically sweet" problems—writing code (that is, computer programs), creating Web sites, building business-to-business communication systems, and content software. The challenges were endless, and the delight in overcoming those challenges was enormous. It could be the best drug in the world and more satisfying than eating a decent meal or sleeping as salve to the soul. A king-of-the-world feeling.

If this high-tech generation were robots, well and good. After all, a robot, too, is self-centered, designed to accomplish certain goals, and in achieving those goals, programmed to work in new similar goals twenty-four hours a day. And these information technologists were prepared to—and often did— work nearly twenty-four hours a day if a work project demanded it, as projects often did.

Of course, the flaw in this comparison is the fact that these twenty-somethings were not robots. They were human beings, which meant that men and women are many-faceted creatures whose happiness requires the fulfillment—or the quest for fulfillment—of connectedness to other human beings, the connectedness of love, caring, compassion, relationships to family, friends, and children.

How best to spend one's life? Two characters in Andre Malraux's famous novel, *Man's Hope*, ask. And the answer is: By transmuting as wide a range of experience as possible into conscious thought. In other words, to delimit oneself is to delimit one's humanity. And without knowing it, that's exactly what happened to these information technologists when they traveled to what they thought was Eldorado, disguised as Silicon Valley, from their home towns in other states and countries. All had been high achievers, but only in the technical fields they graduated from. Their sense of who they were when they graduated was the label they attached to themselves: they were their MBA or Ph.D., but little else. Most of them hadn't a clue as to how to attain and nourish a good interpersonal relationship. Only the feelings for their specific jobs were what they connected with. The joy of a day in the country with a loved one, the expectation of the birth of a first child, the excitement of working for a political cause they believed in, the spending of time in a community service like aid to the homeless or Big Brother or Sister organizations, the curiosity to stretch one's interests by developing new hobbies in non-technical fields, the desire to extend gratuitous help to others.

With few exceptions, these twenty-somethings had boxed their lives into a tiny corner of experience: they were only high-tech workers who thought that that title was the key to the universe. Most of them had chosen a high-tech career because they felt it was a safe place to exist. In school they had been called geeks and nerds by students in the social sciences and humanities, who were adept at connecting their lives with other human beings. But the "geeks or nerds," labeled as such by their fellow classmates, felt demeaned as outcasts. They were "weird," which was the operative word the more socially oriented students used to describe them. It seemed strange to others that a man or a woman could spend twenty hours a day

locked in a room surrounded by computers and their periph-
erals and feel happy in working with them to the exclusion of
even eating or sleeping well.

There were many reasons why this kind of self-isolation felt
safe and comfortable in contrast to associating with the world
where people connected with each other: dances felt scary, pol-
itics boring, dating threatening. There were reasons like living
in an alcoholic family or with violence-prone parents, being at
the center of a bitter divorce, or feeling unloved by parents who
were self-centered. Best to isolate oneself and get away from
threatening interpersonal relationships. Machines were the
solution: they didn't talk back, neglect you, abuse you, or
demean you. Since you were in control of the machines instead
of them controlling you, they were safe. Isolation was safe;
nobody could harm you.

We have seen that even in good economic times this feeling
of "safety-in-isolation" was beginning to pull on many twenty-
somethings and became a major reason why many of them
came to Pat and me for psychotherapeutic help. Now, however,
when the economic good times had become a thing of the
past, an identity crisis, in addition to a feeling of malaise, began
to overwhelm many of them. For now that the jobs they held
had disappeared and new ones in the field of their expertise
were nonexistent, who were they? For many, it became a time
of believing they were personal failures (after all, they always
had been high achievers since they were children, praised for
their technical expertise that no one seemed to want now). For
others it became a time of blaming the victim: they were fired
because of grudges against them by fellow workers, and there

were those who said it would all turn around for the better tomorrow (although tomorrow never came). Then there were the angry, bitter ones who felt their companies had betrayed them, telling them they would never be fired and then firing them, often through an e-mail, with one-day's notice and no severance pay. Possessed by anger and a thirst for vengeance, they would seek out lawyers to sue their former employers for damages inflicted on them.

All of these avenues of anguish were ways they were protecting their long-developed self-images that they were valuable human beings solely because they were high-tech careerists. For hadn't they had the last laugh on all of their former high-school and college classmates who thought they were weird? They had been making $75,000 plus a year, with options added, while their former classmates were lucky to make $30,000. They could flaunt their economic superiority as payback time for their being shunned in earlier years by their classmates.

But now their self-esteem was under attack. Many felt their stomachs turning into empty pits of fear. If they had no career identity, then they were nothing! That feeling is the equivalent of death. It's the equivalent of the robber, who stripped of his gun, feels like he is nothing, which is the scariest feeling in the world. If a gun makes the robber a "somebody," so too his or her career made the high-tech twenty-something a "somebody."

Fortunately, this needn't be the case. The loss of jobs in Silicon Valley could be a wake-up call for the twenty-somethings who feared that their careers were the only definition of who they were as human beings. Many of them have been viewing their careers in a new light and are reevaluating their value structure they have lived by and are finding it was wanting. They are beginning to expand their own sense of themselves: some have moved out of high-tech careers and reinvented themselves as lawyers, teachers, political activists, writers, environmentalists—and even actors!

These are the people who have viewed their crises in the Chinese sense of the word: crises as a time of disaster *or* opportunity. These former high-tech employees seized their crises as an opportunity to expand their lives, rather than remain in a self-imposed constructed box.

Others, however, have remained vengeful, self-pitying, and victimized, waiting for payback time to happen—all in the service of protecting their old, outmoded image of believing they are valuable human beings solely because of their high-tech careers.

EATING UP THE WORLD CAUSES INDIGESTION

What about the Silicon Valley "icons," the "old money:" CEOs who became billionaires in companies like Cisco, Intel, Netscape, AMD (Advanced Micro Dynamics), Apple, Hewlett-Packard, and Microsoft, which were stable companies long before 1995 and are not likely to disappear in the near future? Accustomed to lasting economic power, rather than being two-year disappearing acts, what is the quality of their lives as human beings? We know, of course, the height of their entrepreneurial success, since they have been quick to affirm it in their lavish lifestyles which were glorified repeatedly in the media. A typical example is this awe-struck report on Oracle CEO Larry Ellison's new home by a commentator for the *Silicon Valley Business Report* on March 30, 2001 (a popular show on KICU TV, located in San Jose, Silicon Valley's center):

> If you haven't gotten a chance to peek at the little bungalow Larry Ellison is building in Woodside, we have a birds-eye-view. Despite the fact that Oracle's stock is down to just about 15 dollars a share, its Chairman and CEO is building a 23-acre, 80-million dollar home. The Woodside estate is described as an imperial-style Japanese palace. It will have 10 buildings, including 2 main houses and a Japanese tea house. Some neighbors complain that it's over scale and inappropriate for the area, but others argue that Ellison could have divided the property into 3-acre parcels, which would have caused congestion.

CEOs themselves and/or their media representatives are very verbal about their corporate business achievements (their catchword is "growth"—and more growth: growth in investment, growth in earnings with each quarterly period's growth greater than the last). *Fortune* and dozens of other business magazines will describe in loving detail many CEOs' business deals and bottom-line triumphs, but never anything about who they are as human beings apart from their CEO status. If they say anything else about a CEO, apart from his or her vaunted business expertise, it will be an incidental throwaway line that, oh yes, he or she is married or divorced or single. And if he or she has any children, information about them is minimal.

In other words, what you see about these CEOs is what you *don't* get. A TV interview with them is an invitation for the audience to fall asleep, that is if you are interested in more than just the towering wealth these individuals have garnered for themselves. On the other hand, you may find them fascinating if you are solely interested in making money to the exclusion of everything else.

It has been Pat's and my experience in interviewing many of them that they are interesting for what they *don't* reveal about themselves. And often it's because there is nothing to reveal except their own sense of being a CEO. In James Joyce's perceptive phrase, they seem to be living "a little distant" from themselves. It is as if they surround themselves with a moat: the message they send out is: Don't get too close to me, accept the image I wish to present to you, which is that of an all-powerful god who has shaped his or her destiny alone.

There is the story by that superb writer, Max Beerbohm, called "The Happy Hypocrite," in which the protagonist pretends to be a hypocrite in order to attain his goals. But by the end of the story, he actually has become the hypocrite he had pretended to be. In a similar way, viewing CEOs from our psychological perspective, the CEOs themselves appear to have become the media image they present to the world: one dimensional men and women, rather than well-rounded human beings.

There is a marvelous scene in the classic 1941 film *Citizen Kane* where a reporter interviews a millionaire in his giant, gloomy mansion of a house (a million dollars in 1941 was like a billion today!). The lonely looking, sad-faced millionaire says, when the reporter suggests he must be content with his life: "Anyone can become a millionaire if that's all you want to be, a millionaire." Obsession is the driving force.

Philip Gourevitch, an outstanding reporter who has interviewed many of these high-stake business players, has observed that "You have to have a very grand view of your own capacities and judging by the types who make it very big, a touch of megalomania is probably to be expected." (*New Yorker* Web site, July 7, 2001).

CEOs present themselves to the public as protected castles, invulnerable to any onslaught against them. Consequently whatever personal problems they may have (such as fear of failure, sexual disorders, loveless marriages, neglected children, bitter divorces, barren interpersonal relationships, alcoholism and/or other drug problems) are closely guarded secrets. They will not seek psychotherapeutic counseling for their vulnerabilities and character disorders (such as manic-depression, paranoia, obsessive-compulsive behavior, and anxiety attacks) since they would have to acknowledge to themselves that in addition to being giants of wealth they are also fallible human beings, like the rest of the human race. Examples of how CEOs have tied their total identity to their economic position in the companies they dominate are exhibited when their retirement is announced. How they react to their retirement defines how they view themselves as total persons. The answer is that they usually reveal that they were their careers. Leave a career and what remains? Jerry Sanders, the AMD (Advanced Macro Devices) billionaire CEO is set to retire in April 2002, and reveals in an October 1, 2001, *Mercury News* interview a typical answer to that question: He was asked, now that he will retire, how will he fill his time? He replied: "There's a certain amount of absolute terror when I think what it will be like." This kind of honesty is rarely acknowledged to the media by retiring CEOs, since it demonstrates a vulnerability they are loathe to admit to. Accustomed to being in charge of their world when they are CEOs, they fear the void that retirement will create, since it is a world that is neither predictable nor controllable.

Their "grand view of themselves" and their "touch of megalomania" that Philip Gourevitch reported is evidenced in the philosophies they espouse. For they will talk in general about the values and principles that guide them quite openly, and in doing so reveal out of unawareness the serious limitations their values impose on their own humanity. Many of them like to think of themselves as philosophers to the world—and the philosophy that is most frequently espoused is the philosophy of Ayn Rand. Given their nature, it's hardly surprising that they would espouse a philosophy that exalts selfishness as the most valuable quality in life. In *Business 2.0* magazine's August 2001 issue, Christopher Hitchens reported on this predilections of the movers and shakers in Silicon Valley for this philosophy:

> We find Randianism rampant among Larry Ellison of Oracle, T. J. Rodgers of Cypress Semiconductor, Eric Greenberg (formerly of Scient and Viant), and Kevin O'Connor of DoubleClick, whose son's middle name is Rand. Then there are the Rand brands—named for outfits

or characters in her fiction—such as Rearden Steel, started by WebTV's Steve Perlman, and Galt Technologies (now part of Quicken.com), which was co-founded by Robert Frasca before he moved on to Lycos and Internet Venture Works. Rand Interactive speaks for itself. True Randians call themselves "objectivists," after the pitiless and materialistic philosophy evolved by their guru, which promulgated thoughts and actions by which man must abide to live a proper life. The basic principles of objectivism are objective reality (facts are facts), reason (man's sole means of obtaining knowledge), self-interest (happiness as the highest moral pursuit), and laissez-faire capitalism. In short, you owe it to yourself to be happy. Later this fall, business publisher Texere releases *Ayn Rand and Business*, a primer that may come in handy for current acolytes who neglected to take an objectivist immersion course. For example, Greenberg told the *San Francisco Business Times* in March that a main interest and goal was philanthropy. That may be a commonly admired virtue; Rand, however, had cold contempt for all ideas of charity and compassion and meant every word she wrote in her book *The Virtue of Selfishness*. But there it was on Greenberg's "favorites" list: "Most influential book: *Atlas Shrugged*."

Seeing the world as a jungle in which only the fittest survive (and they themselves define who the fittest are!) is also manifested in the books many of these CEOs have written or ghostwritten. Here is how Charles H. Ferguson, founder of Vermeer Technologies, describes business life at the upper echelon in Silicon Valley:

"It was like palace intrigue in fifteenth-century Venice or

perhaps present–day Iraq. You just assume that anyone might try and kill you at any time, that everyone has a hidden dagger and that dagger is probably poisoned. Your food taster becomes your best friend, especially if he has a food taster too, and even more so if they have to taste your food after you've eaten it as well as before. . . ."

Appropriately enough, Mr. Ferguson titled his book *High Stakes, No Prisoners* (Random House, 1999).

Proof that this attitude was no exception in Silicon Valley's consciousness is afforded by the publication of Andy Grove's best selling book, titled *Only The Paranoid Survive*, published in the same year, also by Random House. Mr. Grove is the leader of Intel, the world's largest chip maker, and is admired for his ability as a businessman. However, his book title, which he proudly acknowledges, is not a guideline for a happy balanced life. The culture of paranoia is not something to be proud of, but rather to be viewed more as a serious psychological character disturbance than a badge of honor. It is the Ayn Rand point of view, that being number one is the only thing that counts in business, and by extension in life. Compassion, empathy, and concern for the human condition, become sentimental side issues not even worthy of consideration in this ice-cold view of the world. Consider Mr. Grove's comment about his employees:

> If you are an employee, sooner or later you will be affected by a strategic inflection point. Who knows what your job will look like after cataclysmic change sweeps through your industry and engulfs the company you work for? Who

knows if your job will even exist and, frankly, who will care besides you?

Until very recently, if you went to work at an established company, you could assume that your job would last the rest of your working life. But when companies no longer have lifelong careers themselves, how can they provide one for their employees? . . .

The sad news is, nobody owes you a career. Your career is literally your business. You own it as a sole proprietor. You have one employee: yourself. You are in competition with millions of similar businesses: millions of other employees all over the world. You need to accept ownership of your career, your skills and the timing of your moves. It is your responsibility to protect this personal business of yours from harm and to position it to benefit from the changes in the environment. Nobody else can do that for you. . . .

This candid statement exposes the media hype that prevailed during the period in which this book was written that proclaimed Silicon Valley employees were a precious corporation resource and that "we're all family" was what the "new economy" was all about. Of course, when the downsizing bloodletting began in the Valley in 2000, the "we're all family" mantra was transformed into the Andy Grove statement that a company's loyalty to its employees was worth exactly nothing.

The latest philosopher-king of the high-tech world, plus the world at large, is Jack Welch, former Chairman and CEO of General Electric from 1981 to 2001. His new book is titled *Straight from the Gut* (Warner Books, 2001). His book is a typical series of business advice, clichés (mixed with autobiography) found in hundreds of other business books. But because of his status as one of the wealthiest men in the world, he has received the adorations of the business and media communities. They overlook the fact that Jeffrey A. Kramer, vice president and publisher of the business books division of McGraw-Hill, points out in his *New York Times* article (09/10/01) about Welch:

"He was known as "Neutron Jack" for firing 100,000 workers and he still lets the "bottom 10 percent" of G.E.'s workforce go every year. . . . and he earned the enmity of environmentalists by opposing a dredging plan to clean up PCBs that G.E. dumped into the Hudson River decades ago. . . ."

Perhaps the best self-definition of the quality of a CEO's life is Jack Welch's own definition as to what is needed to be a successful business leader: "You've got to be on the lunatic fringe," he once said.

The above chapter was written by us just before September 11, 2001, happened. There is the story of the flea floating confidently down the river until he sees a closed drawbridge in front of him.

"Open the bridge! Open the bridge!" he cries in panic. "Open the bridge, can't you see I have an erection!"

For Silicon Valley leaders, and everyone else in this world afflicted with megalomania, September 11, 2001, transformed anyone wearing the arrogance of power into that naked flea who had to learn a lesson in humility.

ARCHETYPES OF
SILICON VALLEY

I t's been said that everything has changed in our society since September 11, 2001. Indeed, the world situation and our place in it has changed: we are now a vulnerable nation that has sounded the alarm bell that we are all one world. The trendy term "globalization" has taken on a new, more menacing meaning. Globalization can and does equal danger.

Silicon Valley has always been obsessed with "the next new thing": create a new peripheral, a new service, a new need, a new sequel to the silicon chip. This was the survival mantra in an industry where a company competes successfully or dies, and discovering the "next new thing" was always considered the key to survival.

However, since September 11, the next new thing has turned out to be not a product or service, but fear itself. Of course, fear was always the underlying force that drove Silicon Valley's competitive companies. Number one or die! was what we heard with boring frequency from the corporate spokesmen we visited. Now, however, less is spoken about new innovations as the driving force for maintaining a successful corporation. Instead, fear itself is the driving force. Now survival takes center stage in the Silicon Valley economy. Consolidation and nurturing resources, rather than the over-expansive drive in mad pursuit of the next new thing that contributed to the Valley's recession, now dominate in the industry.

But has human nature changed since September 11? It's not that easy to change at fundamental levels: existential change is difficult to achieve. We all know of the superficial change that can occur in a person who has had a life-threatening heart attack or a cancer. Suddenly that person, possessed by fear, begins to make a bargain with God in his or her mind: "I'll be a better person if you let me live. I won't be selfish, or inconsiderate; I'll be kinder and more generous."

That's the bargain made in panic. However, when that person's heart heals or the cancer disappears, the man or woman rarely changes. For there is comfort in the familiar. So if one lived a lifetime being self-centered, with rare exceptions, one will remain self-centered. A person, to whom money was everything, will plea bargain with God and say there are more important things than money and that he or she is now a changed person. When the illness passes, the money obsession returns.

In our counseling practice, we often see people who have severe personal problems, but who think there is an instant cure in three or four counseling sessions. We always try to ascertain the degree of commitment of such a person to change: If that person is willing to take a long inward journey to discover and overcome the self-defeating behavior that has made life miserable, that person can truly experience existential change. Otherwise, we tell clients unwilling to take that journey, that perhaps they are not ready for such counseling, but may be later on.

On the other hand, September 11 affords each and every one of us the rare, most-valuable, opportunity to reflect on how we wish to live our lives in this time of profound uncertainty and menace.

It would be foolhardy to believe the human race can change its face overnight. But our triumph over the fear and terror inherent in this world can begin to happen if we center our attention, and practice the meaning of John Donne's immortal injunction:

> No man is an island entire of itself. . . . Any man's death diminishes me, because I am involved in mankind. And therefore never seem to know for whom the bell tolls. It tolls for thee.

The way we live our lives is what John Donne is writing about. And September 11 is a flashing neon sign for us to pay attention to the fact that that the time of each of our lives may be shorter than we think. In our last chapter, we have reported on the quality of the lives of the people living in Silicon Valley. In this section we will share a series of archetypes of people who live and work in Silicon Valley. The dictionary defines an "archetype" as an original model or type after which similar things are patterned. The following five archetypes are people we have interviewed and counseled whose attitudes and lifestyles reflect those of thousands of similar Silicon Valley men and women. Their experiences define the ways they value their lives:

1. The Engineer Who Discovered He Wanted a Life
2. The CEO with No Identity
3. Why Not Me?
4. Two Can Be Lonelier Than One
5. The Dispossessed

1. THE ENGINEER WHO DISCOVERED HE WANTED A LIFE

Jerry is a top-of-the-line information technologist, a troubleshooter, responsible for keeping the system of over three hundred computers functioning well in the financial corporation he works for. He is thirty-two years old, has an MBA, and worked for the same company for seven years, making him an "old-timer" in the field.

Initially he liked his job: solving technical problems came naturally to him. The challenge of making computers function well had always been a triumph for him.

He grew up in Chicago before moving to Palo Alto in Silicon Valley. His father had worked as a salesman in the same office supply firm for thirty years and retired at the time Jerry left for the West Coast. His parents were then in good health in their sixties when he left Chicago. They took pride in him as an only child, and glowed in his 3.9 college grade point average.

Jerry's love for computers was an early infatuation. He was more comfortable with creating, mastering, and solving physical problems. Taking apart a computer and putting it back together again was a sense of passionate excitement to him. The joy of building his first transistor radio when he was seven was a feeling he never forgot. He had little interest in dating or in social functions since writing computer code, Web design, and accessing esoteric communication systems were his form of entertainment and socializing. For as an only child he learned early in life to take care of himself, and to entertain himself, and to create challenges for himself since he had no brothers or sisters to resonate with. His fellow students saw him as a shy, retiring, pleasant-faced young man who never felt very comfortable in any campus group.

Jerry had dated sporadically. It seemed the women who were attracted to his pleasant, nonthreatening demeanor were soon bored with him since he only seemed to exhibit excitement when he talked about the latest computer innovation or application.

He was different from his parents. His father was gregarious and liked to have his friends over to the house; his mother enjoyed playing the piano. Neither had any interest in the technical engineering field that so engrossed their son. They viewed him with proud amusement and wondered where in the world did he get the kind of interests that were so foreign to them. It was not a condescending attitude on their part. In fact they fully supported Jerry in his choice of a career. He was their precious child, their high-achieving son.

Jerry agreed with them. He saw himself as exceptional

(after all, most of his fellow classmates didn't know a damn thing about computers!). He saw his career as the most important thing in life—the key to all the other prerequisites of existence. A family, maybe children, good friends? Later, later, was his way of thinking. Spending eighteen hours a day on his job, sometimes working seven days a week, precluded other forms of activity, particularly since the work was so enjoyable, and his pay ($100,000 a year) made him feel superior to the fellow college graduates he knew who were unhappy in $30,000-a-year jobs. The company's perks (good health and dental benefits, a splendid pool, a fine gym) eliminated the need for him to go outside the company's premises to fulfill his needs. The company said, "We're all a family here" and Jerry believed it—after all, look at how the firm took care of him.

But when he was five years into his career, he began to feel twinges of dissatisfaction. It had to do with the fact that he had just turned thirty. He returned to Chicago to celebrate his birthday with his parents and their friends. He noticed he had no friends to invite to his party, and he felt a sense of inexplicable loss. He was beginning to feel the need for "something more" in his life. His parents seemed so happy with their friends. Even though his parents had always been concerned about making ends meet (their income was middle-class modest), they always seemed to be happy with their lot in life. Jerry had always felt that since he was going to make much more money than they did—and he did in Silicon Valley!—that he would be far happier than they were, since he had often heard their complaints about money.

He also noticed, during that one-week visit home, that he

liked the family routine his parents always had practiced: family dinners every night and the good talk with everyone at that time about how the day went. His dad and mother loved the theater and would take him, ever since he was nine, to see a new play or musical. And this time, too—at the age of thirty—he went with them to see a musical that delighted all of them. A sad smile came over Jerry's face when he remembered that he never went to the theater once in the five years he lived in Silicon Valley. Yet San Francisco was a wonderful theater town, and only forty-five minutes from where he lived. He also felt poignantly that the small house his family had always lived in was a place of comfort, rather than the constraint he thought it to be when he left for Silicon Valley. But in the Valley he was living in a small apartment ($4,000 a month for one bedroom!), which he used as a place to hang his clothes, eat an occasional dinner, and sleep for five or six hours. Suddenly, his old family house looked larger-than-life and had a welcome-home feeling to it.

And what about family life? His parents were rooted in their community. They went to dances and block parties and did volunteer work for the cancer society.

Jerry experienced an epiphany—maybe he was ready for family life; maybe he was ready to date, expand his interests and even consider having children. Maybe being a parent was a greater challenge than an information technology career.

When he returned to Silicon Valley, it looked changed. Of course, it wasn't, but he himself was changing. It all began to seem so shallow and superficial. The belief that engineers were going to change the world for the better was an illusion. Jerry smiled when he thought how naive he had been. It was all a money game. Everybody was competing with everybody else. Being number one meant everything, and yet there was so much discontent and unhappiness in the Valley. Jerry knew of three associates whose wives divorced them because they felt they were in an impossible triangle. The common complaints from their wives were: "You love your job and the company more than you love me. You never see our children. Not once did you go to a little league game. You're a rotten excuse for a parent. I need a life and I'm going to get it, but it won't be with you any longer!" His three associates earned six figures, had comfortable houses, yet were miserable, embittered, vengeful guys. Everything seemed out of balance to Jerry since his return from Chicago.

Thirty is frequently a turning point for men and women—a time for reevaluating what they want out of life. Usually, their career had assumed top priority during their twenties. Now it became a been-there-done-that affair. What's next? became a gnawing question for people like Jerry.

Nothing changed outwardly for Jerry during the next year after his return from Chicago. But inwardly his dissatisfaction with his lifestyle was growing. It reached a boiling point when, seven years into his career, he received a frantic phone call from his mother. His father had a massive heart attack and had just come home from intensive care. Modern medicine gave him the possibility of living a few more good years of life.

Jerry left for Chicago the next day. Homecoming was a happier affair than he expected it to be since his father was recovering nicely. But the uncertainty of life sounded in his mind like thunder. What would he want to be remembered for if he should suddenly die? Jerry asked himself. He did not like his answer, which was: "Jerry was a man capable of fixing computers!" On the other hand, his father would leave a legacy of family, friends, and community accomplishment, even though he never made enough money for what Jerry considered would be real happiness.

But there was no real happiness in his Silicon Valley life! He now experienced that fact viscerally. He found it hard to look at himself in the mirror.

This time when Jerry left Chicago, it would be a time to make other plans for his life. He would use his engineering skills not as a substitute for other aspects of life, but as an enhancement of them. He had earned enough to start a second career. He would return to school, get a Ph.D., and teach. He would have regular hours, in a college environment, that would enable him to expand his interests, meet new associates, and begin to date. Money of course is something to be concerned about, but not to be obsessed about like he once was.

When we last heard from Jerry, he had quit his job, left Silicon Valley, and returned to Chicago. He is now in the process of getting his Ph.D. and tells us he is far happier than he ever was in the Valley. His last letter, in part, told us: "In a strange way, I find I'm very similar to my parents. When I left to take a job in Silicon Valley, I thought I was a bit ashamed of them because of their low-income status. But I now find that I'm very proud of them. They are the success, I was the failure! I'm dating a classmate now and she's taking me to dances and political rallies. The surprise is I like them! I'm beginning to place my experience in Silicon Valley in perspective. Silicon Valley is like a smorgasbord. You get what you want to put into it—and if it's not your piece of cake then leave it for the next person to indulge in.

"I'm fortunate I left before the big economic recession but I'm lucky I'm not what I once was—a guy who put all of his eggs in the company basket."

2. THE CEO WITH NO IDENTITY

Paul had a talent for mathematics. If playing a piano comes naturally for some people, becoming a mathematics whiz was second nature to Paul. He was thirty-five, and he had utilized his mathematic abilities to become the CEO of a well-known debt-management firm highly regarded and in demand by major Silicon Valley corporations.

You would never guess his origins from the manner in which he presented himself to the public at large. His grooming smelled of money: the London tailoring, the modest-elegant tie and shirt, the fine-fabric suit (usually charcoal gray or black), and the shoes that hid their excessive price. He was an imposing six foot one inch tall, not a bit overweight, and had a cultivated voice that matched the in charge aura he generated when he made an entrance into a room.

He was born dirt poor in a run-down section of St. Louis. His father was a construction worker in a field noted for its erratic employment, and his mother was the complaining wife whose occupation was the raising of three children. Paul was the oldest, and Terry and Jane were his siblings.

Paul's father, Harry, was a raging alcoholic who had ravishing good looks and an intelligence that belied his lack of even a completed high school education. His mother, Betty, came from a middle-class family that told her she would be making a serious mistake if she married Harry. She rarely lost an opportunity to remind Harry of that fact.

Harry was a perpetual drunk. He rarely had a sober moment. The struggle over household finances was endless since most of the money went to the neighborhood bar. Paul was often told by his mother to go to the bar and pick up his father before he spent all his money.

Harry was an angry drunk who used his fist to solve family problems. Paul learned to survive by keeping out of Harry's way when he came home. "I was quiet as wallpaper because if I wasn't I would be bloodied up by him," he told us, and continued:

My brother was the shit disturber. He was very foolhardy. He would disagree with my dad and then get hit with my dad's fist or whipped with a heavy leather strap. He particularly got it when my dad would be slapping my mother around and Terry would scream at him to stop it. . . .

I said I was like wallpaper, since I saw what happened if you made my father angry, which was over practically anything. I would peek through the door in the next room when my dad would maul my brother. I thought my brother wasn't brave, he was foolhardy! I wrote my father off early in my life and couldn't count on my mother to protect me. I trusted no one except myself, and you know, that's the best advice my family ever gave me.

My dad always put me down. It was as if the only way he could convince you he was worth something was by putting the other person down. Since he never completed high school, he would always put education down. I was on the thin, fragile side, so he would say he was better than I was by showing his arm muscles and tough body.

But I knew I was good at school, so I didn't believe him. I would get A's in everything, and a math teacher of mine, Mr. Evers, encouraged my interest in mathematics and said I could make a career out of it if I wished to. I could become a statistician, according to him. That's a laugh. He saw I had a low career expectation for myself and thought that I would think becoming a statistician would be a big deal. I was sixteen at the time and was worried about ever finding any kind of permanent job, let alone a mathematics job. Mr. Evers is dead now, but he would be proud of me, I think, because he did help me get a college scholarship based on my mathematics ability. I did very well and got a fellowship and earned an MBA.

It was getting that college scholarship that saved my life. The school was five hundred miles from home, so I could escape that snake pit when I graduated from high school. My father told me I was a fool to go to college; making money in construction was his idea of a good life. My brother, no matter how much he had fought with my father, became exactly like him. He stayed home and also became a construction worker and a full-time alcoholic. He would repeat on his own wife the physical punishment my dad had given my mother. My sister married, stayed in St. Louis, and became a bookkeeper. She has a decent husband, but worries about her son who is retarded.

I was the only one that escaped! My mother never spoke

to me about it, but her attitude told me she knew I was right to leave, but that she felt less safe with her children gone, and she alone had to face that bastard husband everyday. I think I loved my mother, but I was very angry at her. I felt she was a wimp, that she should have left my dad years ago and taken me and my brother and sister away. I could never count on her to protect me from my dad.

I didn't realize it when I was a kid, but growing up in my kind of crazy family was good training for my becoming a good CEO! First of all it made me trust no one except myself. Second, it made me become a self-starter. I couldn't count on anyone or anything except my own ability to make good things happen. I could take risks because I felt I never had anything to lose. I could never trust anyone to be there for me when I needed someone. After all, relying on my mom and dad was like falling off a roof. My family taught me to be fiercely competitive. I was going to be number one in whatever I did in life and show my dad how wrong he was about me. I was a very close observer of people. My dad taught me how to be wary of even the closest person in my life. I had to pick up all sorts of nuances of my dad's behavior to make sure I could avoid angering him or getting another whack on my head. I had to stuff my feelings. Like I said, I was wallpaper at home—always wary, always observant to avoid bad things from happening to me.

Since I couldn't convince my family of my value and worth, I felt I was going to convince the world about how capable I was. And that meant making money, tons of money. Don't let anyone bullshit you; money is power, all the rest is window dressing. . . . Oh yes, if you're not born rich, you have to work your ass off, every day, all the time to become a success. Always be on guard, never take anything for granted. Whenever I thought my father would stop drinking forever, he would always disappoint me by going off the wagon the very next day after he said he would stop.

I learned to survive in that hellhole of a family by becoming all things to all people. Life is a game to me. Always look pleasant, look interested, look compassionate, and act as if you always wanted to be helpful. In my business it's important to win people's trust. I always show I *care* about their problems. I'm in a highly competitive industry, so it's important to win the trust of my clients to get the big-time deals that make my company one of the top five in its field.

I never forget what a high-level international financier who was invited to speak in my money and banking class at college said. He said, "At the high levels of corporation business there are no ethics and there are no rules. It's the survival of the fittest with all the gloves off."

I can remember how shocked most of the class was when he said that. It violated all the fancy words in all of their textbooks. But it's true, and the reason why I knew it was true was because that's the way my family taught me to grow up! What he said was no different from what Balzac, one of my favorite writers wrote. He said, "Behind every great fortune is a great crime." That's a little extreme, but that's the way the world works.

Paul was an extraordinary expert in his business. He could enthrall clients with his dramatic graphic presentation of complicated data; he could translate the esoteric jargon of business mathematics into homely phrases that would flatter listeners. His clients relied on his economic predictions, which he presented with the flair of authoritative wisdom. Those predictions were more successful than his competitors. "But it's all smoke and mirrors," he laughed when he told us of his presentations. "Nobody in the world knows enough to predict economic developments with any degree of accuracy. But in my business you have to give the impression of scientific accuracy. So I have the mumbo jumbo of colored charts, statistical projects, and background technical reports. But I'm smart, so my educated guesses turn out to be right. I then can brag to the corporate media that my company used the cutting-edge mathematical findings to keep us ahead of our competitors. As I said, it's all a game."

Paul believed in nothing and believed that was the foundation of his success. But he was one of the unhappiest persons we ever met. His permanent facade of hail-fellow-well-met had crumbled, and a dark cloud of sadness took its place.

Why did he come to see us? The things that were most important to him—his wealth (he was close to being a billionaire) and his CEO status that he felt affirmed his worth as a person—had lost their value to him. And he was beginning to make wrong business decisions.

We counseled him over a period of three months, and then he disappeared from our lives as suddenly as he entered them. He was a man who had lost his sense of self because he *never* knew who he was. What triggered his loss was the death of his father. Two months before he saw us, he had been informed of his father's death by his brother. Of course it was alcohol—his liver was practically nonexistent, and he was only sixty-four. Paul didn't attend his father's funeral. As usual he sent some money to his mother and felt that ended his relationship with his father. Of course, it didn't. He still was carrying his father with him. He was an ever present elephant in Paul's bedroom. Paul began to have nightmares. One dream repeated itself: In the dream his father was very much alive and very drunk. He caught Paul (who was twelve years old) hiding a dollar bill out of the money he had earned from selling newspapers. "That's my money," his father shouted and ran after Paul with a knife in his hand. His father tripped and fell on his knife, which killed him.

It turned out that Paul's compulsive drive to "prove himself" was an unconscious, anguished desire to win the approval of his father. His parents had created him and half of that creation was his father's doing. And if he felt—as he did—that his father thought he was worthless, then that would mean at the unconscious level half of Paul was worthless. And since his mother never took issue with his father, that meant all of him was worthless! Paul had to look for validation of his value as a human being *outside* of his family. But that validation could never be a substitute for his parents' validation. All of his life, Paul was living with a bottomless pit inside his psyche: he was looking for his sense of self in all the wrong places. Money and power can enhance one's feeling of self-worth, but it can also diminish it if it becomes a substitute for his parental love.

Paul saw us half a dozen times in those three months, then said he was feeling in control again and terminated his counseling sessions. The process of denial was at work. A certain amount of self-healing took place, but he was unwilling to face the contradictions in his psyche—the signal of being worthless as a person and wealthy as an "icon." He had avoided confronting many other demons in his unconscious: there was his fear of commitment—his inability to trust any woman enough to maintain an ongoing, monogamous relationship (expensive one-night-stand call girls was his definition of a relationship). There was the need to always be in control of every situation. There was his fear of failure, the need to be number one in every situation. There was his lack of empathy, his inexplicable rages, and the desire for vengeance against anyone who slighted him.

It's been three years since we last saw him. We hear he is still a top CEO, still wealthy, still driven. And he is still the loner waiting for the next depressive incident that could happen around the next corner in his life.

3. WHY NOT ME?

He did not like to be called "Nick." His name was "Nicholas," and it was a matter of pride, a matter of dignity, that he insisted on responding only to that name. His parents were dirt poor. They came from a small town in Greece where they labored in barren soil to eke out a margined living. They came with their three children back in 1968 when Nicholas was fifteen. He was the eldest; his two younger sisters were eight and eleven.

The family wound up in San Jose, which still exhibited a small-town aura rather than the sophisticated Silicon Valley city it was to become. His parents created a produce business and made a modest living that enabled them to send their children to school. Like any loving parents, they wanted their children to achieve better lives than they had experienced, and education was the key to that goal.

The children flourished in the new environment. All three graduated from college—one of the two sisters became an office manager, the other a clinical dietician. Nicholas liked working in the produce industry and became the owner of a small catering business that grew as Silicon Valley grew. Hard work was his inheritance, so working seven days a week, seventeen hours a day when necessary, was no burden to him. It was just the way of the world, so why complain.

It was important to him, coming from an immigrant family background, to believe he must become more American than if he were native born. He made a point of marrying a woman of Wasp extraction and had two children—the boy was named Jonathan, the girl Sally. American names. His wife, Alice, worked with him in the catering business. It was a partnership in which two caring people established a solid foundation for themselves and their children. It all seemed like a standard American Dream story: an immigrant comes to the land of milk and honey, partakes of its riches, and leads a happy life ever after. Nicholas believed in that dream and was one of its most enthusiastic endorsers.

But there was a dark side to the American Dream that revealed itself in the growth of Silicon Valley. On the one hand, America was the land of equal opportunity for everyone. On the other hand, to paraphrase George Orwell, some people are more equal than other people. And in the explosive growth of San Jose (labeled in the eighties and nineties as the heart of Silicon Valley) gold rush fever, an "I'm-richer-than-you-are-therefore-I'm-better-than-you" attitude overcame the egalitarian American impulse.

This virus infected Nicholas. When he started his catering business, he was quite happy with the possibilities it offered. A good living from hard work. During the 1980s that's exactly what happened. He had also developed a talent for buying and selling houses. He would take a run-down property, improve it, and then sell it to buy another. It afforded enough additional income to send his two children to good out-of-town colleges.

There was a catch to this prosperity, however. For Nicholas and his wife it was all work, work, work. And now in his mid-forties (the time was 1998), he was feeling burned out. This was the time when Silicon Valley was awash with instant fortunes. Of course it was all on paper. But that was real, wasn't it? Stock options and start-up stock investments were creating instant millionaires. Let the options accumulate, and don't turn your stock into cash yet because it will double its value if you hold onto it another year.

This buzz was deafening, and the media made Silicon Valley the glamour capitol of the world in 1998. Middle-class people throughout the United States, millions of them, were seduced into playing the stock market, since it was the way to make an instant fortune without working. Why not participate in this largesse? For a person like Nicholas, who was living right where all this presumed torrential wealth was being created, it was an even more seductive situation. He catered to the Silicon

Valley companies where the engineers were flaunting their wealth—their grand houses, their private planes, their glamorous parties. And they were just kids, only twenty-five, twenty-six, twenty-seven! Nicholas could feel their arrogance cutting through him when he serviced their needs. He was treated like a servant by these snot-nosed kids!

Why not me? Nicholas thought. If these kids, fresh out of school, could become millionaires, why not me? Didn't he deserve the kind of money they were making since he had slaved for every penny most of his life?

He began to keep his ears open, ever attentive to the rumors, the tips, the "insider" talk about what stock was going to rise another 100 percent in the next two months, about what new IPO was going to be initiated, and how it would make you instantly wealthy if you bought it the day it was marketed.

It seemed so simple; all you needed was money to invest and lo and behold! elegant retirement for the rest of your life. It would be wonderful for everyone—his wife, his kids, himself, his aged parents.

Of course, Nicholas knew little or nothing about the inner workings of the stock market or venture capitalists or even what the giant Silicon Valley corporations were making. They made stuff for computers or made computers or established Web sites known as places for e-commerce, but all of that was irrelevant detail to him, as it was to millions of other middle-class people like him who yearned to invest and make money. It was like roulette where your number came up every time you played; what the companies made or didn't make was irrelevant.

An opportunity arose for Nicholas where indeed he thought he could become very wealthy and retire for life. Because of the wild increase in real estate values in Silicon Valley, Nicholas was able to sell a home for one million dollars he had purchased years before for only $100,000. He had held onto the property for a dozen years in the intuitive belief that it would increase substantially in value due to the thousands upon thousands of new workers flooding the Valley who were making six-figure incomes and desperately needing places to live, preferably elegant places like the one Nicholas had remodeled and sold to a vice president of a new content-provider start-up.

Nicholas' wife, Alice, was also caught up in the Silicon Valley stock market excitement. After all, it was 1998, the height of boom times in the Valley and all the authoritative media analysts and business magazines said the stock market could only go up, unlike previous times. A stock market crash was impossible. A best-selling book at that time had a title that announced the Dow-Jones Industrial Average (DOW) was bound to reach 36000!

However, Alice was more cautious than Nicholas. She agreed to invest some of the money from the sale of the house, but not all of it. She thought Nicholas agreed to invest $250,000 and keep the rest of that windfall profit in the bank. However, that was not the case. Nicholas felt he would be a fool not to invest the entire million dollars in high-tech Silicon Valley stocks. It would be a once-in-a-lifetime opportunity that he would always regret not taking if he didn't grasp that opportunity now. There was no sense arguing this point with Alice since she was adamant on investing no more than a quarter of their profit. But she certainly would be appreciative once his stock market investments would triple in their value.

Nicholas did not carelessly invest. He tried to acquaint himself with as much knowledge as possible about the stocks that held the best potential for rapid growth. They would all be high-tech Silicon Valley–type stocks of course because that was where the fortunes were to be made. Nicholas' research included talking with business associates in his community, brokers, investment counselors, and he diligently read *Forbes*, *Business Week*, and *Fortune* magazines, along with the *Wall Street Journal*. They all said the same thing: high-tech stocks were the wave of the future, the best investments one could possibly make. So Nicholas picked the stocks that the business community consensus assured the public were the big winners in the stock market lottery. He invested in the heavy-duty Silicon Valley companies—Cisco, Hewlett-Packard, Apple—and in newer start-ups that seemed to have a skyrocket potential— Amazon, Yahoo, Webvan, Petstore, Kozmo, and e-Toys.

It was 1998, the year when high-tech stocks could only go up. Nicholas became addicted to his computer and television stock reports. His first act upon waking every morning was to click on CNBC's *Market Watch* to find out how his stocks were performing. He would read the business pages of his local newspaper, the *San Jose Mercury News* while eating his breakfast (his wife would semihumorously remark, "Hey, remember me, I'm here" as her eyes scanned the earnings figures in the newspaper while he silently had his toast and coffee). He would smile and say, "Our stocks are up again!" When he came home for lunch, he really came home to his computer, checking out the NASDAQ and DOW returns of the current day, since the market already closed down in the East when it was noon in San Jose.

Nicholas' catering business, which was stable and needed little personal attention on his part, took a backseat to his intense emotional involvement with the stock market. As his stocks rose, a euphoric look suffused his face. However, Alice was less than pleased. She saw her relationship with her husband deteriorating. It was as if he were married to the stock market rather than to her. Yes, it was nice that their investment was increasing in value, but it was not so exhilarating to find her husband living in what seemed to be a different world. She valued her relationship to her husband more than the stocks they held, and she began to worry that that relationship seemed to be disappearing. At the end of 1999, she suggested to Nicholas that they sell the stock and pay more attention to each other than to the stock market—a long vacation would be nice. But Nicholas refused. There was still a lot of money to be made in the market; all the brokers, the venture capitalists, and the investment advisers were awash with permanent enthusiasm, and the Valley's corporations' activities had become national front-page news on an almost daily basis. So why not hold onto their investment permanently, was the way Nicholas began to think. Maybe even becoming a billionaire was possible?

But after April 2000, this dream began to turn into a nightmare. The new start-ups that Nicholas had thought were smart investments became dust, instead of building blocks for a fabulous fortune: Webvan, Petstore, Kozmo, Govworks, and e-Toys disappeared into history. And Cisco, Hewlett-Packard, Apple,

Yahoo, and Amazon dropped precipitously in value. By the beginning of 2001, Nicholas' investment had lost 80 percent of its value; the million dollars had shrunk to $200,000. From mid-2000 on, Nicholas' face turned into a sea of worry. He wore his stress, his gloom, and his anxiety like a heavy black overcoat which he put on the moment he woke up every morning. He would turn on *Market Watch* every morning in fear rather than with eager anticipation. He prayed the value of his investment would rise each day, only to experience another downturn. He became so depressed that Alice demanded to know what was happening to him. "I know you well enough to know you're hiding something, what is it?" she asked. This occurred in the beginning of 2001.

Nicholas then reluctantly told Alice that he had invested the entire million dollars in the stock market rather than a quarter of that amount that she had suggested. Alice was furious. She thought that he had agreed to her suggestion, but had betrayed her instead. He tried to defend himself; it was all done for the both of them, not because he wanted to betray her. He thought she had been too naive, too unwise not to know that he could make millions out of their investment. Yes, he was wrong, but things could still turn around. Let's keep the remaining $200,000 in the market so we can recapture what we lost.

Alice coldly rejected this suggestion of his. She demanded he cut his losses and salvage the remaining $200,000.

Nicholas felt his pride was at stake, and he had also convinced himself that by mid-2001 the market would turn around (that was the talk of the experts, talk he wanted to believe, even though the experts were wrong before). More than the loss of money was at stake. It was his dream of a work-free, worry-free future that was being destroyed, something he could not tolerate. Consequently, he refused to agree with Alice. He would stay in the market waiting for the turnaround that was sure to develop. He elicited Alice's reluctant agreement to remain in the market until June 2001. Alice agreed on condition that Nicholas would leave the market forever at that time if the downslide continued.

The market, however, continued to punish Nicholas, and by June his million dollar investment was worth $25,000. Alice wanted Nicholas to say he was going to end his obsession with the market, but he remained silent. Still angry, Alice confronted Nicholas and told him it was time to end what she called his "idiotic" behavior. He refused to do so—he would continue to play the market until he regained the money he had lost, even though he had agreed not to.

Alice was more worried about the relationship with Nicholas that she seemed to be losing rather than the money. What hurt her most was his untrustworthiness. For a second time he had betrayed her. Over money of all things. His affair with money had become worse than if he had had an affair with another woman. That was something that never occurred in their marriage. They had been married for twenty-two years and had two lovely children. It was a good marriage, and it would be absolutely horrible, in Alice's eyes, if it should now break up. However, they were living with pain and alienation and resentment. Before they were a team, but now they were discovering that two people can be lonelier together than if they had lived separately.

The tension between them had become unbearable. Alice wanted to fight for her marriage and put it back on track, so she contacted Pat and me for marriage counseling. Nicholas valued their marriage as much as she did and agreed to joint counseling sessions with us.

The story we related above was what Nicholas and Alice told us about themselves in their counseling sessions with us. They came once a week for six months and celebrated the New Year of 2002 as a couple happier than they had been since the beginning of Nicholas' affair with the stock market three years previously.

As we mentioned in the beginning of this book, Silicon Valley is more than just a geographical area in northern California. It is also a state of mind—a state of mind that lusts after the easy life, wealth without working for it, wealth that defines your important status in life. The odds for attaining that kind of wealth by investing in the stock market are as great as winning a lottery, unless you are an insider who knows how to manipulate the market whether stocks rise or fall. Nicholas was one of millions of men and women who invested in the stock market in the late 1990s (more than half of all American families held stock in that decade) and had little or nothing to show for their effort. These were good people seduced by the dark side of the American Dream. For seduction it was: the media trumpeted that everyone was making a fortune, so you should come aboard the NASDAQ train. But the brokers, venture capitalists, and investment firms that wanted the average person's money were interested in making millions for themselves. They could care less for the average investors, whom they knew would be naive victims of a fickle stock market.

Nicholas cared more for his marriage than he did for trying to recapture the money he had lost. That was the bottom line that made the counseling sessions he and Alice had with us journeys in healing and forgiveness. What Nicholas had to give up was not the loss of a specific sum of money, it was the loss of his dream of a life lived on fabulous, unearned wealth. He never meant to "betray" Alice, but he mistakenly had convinced himself he was acting in the best interests of both of them.

When we asked him what was more important, the stock market or his marriage, he was shocked. It was his marriage, of course. Nicholas was confronted with either reaffirming the basic values of his marriage or pursuing a dream of wealth that led nowhere. His pride was hurt, but it was a small price to pay for renewing his relationship with Alice. Alice in turn forgave him for his "betrayal" of her. She recognized his action for what it was—a temporary aberration that made him search for fool's gold. For foolish it was, since she knew that they would never have been happy if he had made the fortune that so obsessed his life. He would have been married to that money for the rest of his life instead of being married to her.

By the time Nicholas terminated his relationship with his broker and the stock market, he wound up with a return of $110.00. The million dollars had been pipe dream money. But Nicholas and Alice were lucky; they had a strong marriage that triumphed over stock market figures. They also had a business that sustained them, even though it required harder work as a result of the recession. Many millions of other men and women who also engaged in the Silicon Valley stock market dream of fabulous wealth were not as lucky as Nicholas and Alice.

4. TWO CAN BE LONELIER THAN ONE

Harry and Carol came to see Pat and me for marriage counseling. But it soon became evident that what they really wanted was divorce counseling. They were both thirty-five and were highly successful Silicon Valley careerists. Harry was a middle-six-figure consultant for networking and telecom companies; Carol was an in-demand Web-designer specialist whose income exceeded $100,000 annually.

Both earned their incomes. They were very hard workers and took pride in the effort they put into their careers. They avoided their eat-drink-and-be-merry cohorts who viewed Silicon Valley solely as a gold mine to be plundered. They had been college sweethearts and grew up in comfortable homes with professional parents in Chicago. They were programmed to succeed in the economic world, and their parents were pleased with their accomplishments.

They were unencumbered by debt, owned a house in Palo Alto they enjoyed living in, and loved their work. But that was their problem (it also was the tip of the iceberg of more disturbing problems that had been eroding their twelve-year marriage). For the enjoyment of their work—the excitement, the perks, the status, the economic income that enabled them to take expensive vacations in exotic places—created a warm cocoon of self-centeredness for them in which life's serious problems could be sidestepped.

It was this very self-centeredness, that previously afforded them so much personal satisfaction, that was creating the problems that brought them to our counseling center. They had also regarded themselves as being in control of their life events— they could predict and plan and make their future materialize. But now life was making plans for them that they could not control. The self-indulgence that had created contentment for them now became the source of their growing alienation from each other. In their previously well-ordered lives they never factored in the possibility that they would be growing older, and in that process of aging, fundamental changes in attitudes, desires, feelings, hopes, and dreams would take place. At thirty-five, Harry and Carol no longer were the contented couple. It was as if life had given each of them a black eye in the form of a profound disagreement as to how they would move the next stage in their life forward.

Carol wanted to have a child and Harry didn't. A seemingly simple issue that really was very complicated. Three months before Carol and Harry decided to see us for marriage counseling, Harry told us that it felt as if Carol had mugged him when she first told him of her desire to have a child. He thought she must be kidding, and he told her so. He had reminded her of their solemn promise to each other at the time of their marriage that they would never want to have children. They were aware of too many disenchanted couples who were saddled with kids and debt and would have none of that for themselves. Personal fulfillment and personal enjoyment was what they valued in life, and that excluded children.

Given the commitment that not having children was the bedrock of their relationship, it was not surprising that Harry thought Carol was joking and would then discuss more important things like what city to visit on their next vacation. So he waited for her to smile and change the subject. However, her determined look and tight lips did not vanish. "I've given it a lot of thought and feel it's the right time to have a child, and I hope you feel the same way," she told him. She waited for his answer, but Harry, who was never at a loss for words, was so disoriented that he became speechless. Regaining his composure, he said, "I don't understand, you know our agreement, so why change now? Hasn't our life been damn good without kids?"

This wasn't the answer Carol hoped she would get from Harry. She thought he would be as excited and happy as she was about deciding to have a child. "That was so foolish of me," she told us bitterly in their first counseling session with us. "I thought he would sweep me in his arms, give me a kiss, and tell me that's exactly what he wanted also. Instead, after I pushed him for a direct answer, he said, "Hell, no!" that felt as if he had given me a real hard slap to my face. But I thought maybe we needed more time to talk about it, maybe I could change his mind. Fat chance!"

For the next three months, they felt they were living in two separate households. They had always slept in the same bed, but that became painful because they were always thinking of the issue that divided them. Harry moved into a second bedroom and each night overdosed himself on bad television reruns and sizable belts of gin to tranquilize himself to sleep. Carol would silently cry and felt like a ship without a sail alone in their king-size bed.

Both of them remained adamant in their positions on the baby subject. Words like, "You're mean," "selfish," "a cruel son-of-a-bitch" were the verbal bullets Carol shot at Harry. And Harry would shout the word "Betrayal" over and over again, with "You're an untrustworthy bitch, I should have known that before I married you!" added on to his attack.

Their bitter words were really Halloween masks, with each of them scaring the other to death. For fear was the root cause of their alienation from each other. Without their realizing it, they had taken each other for granted. They assumed they knew each other because they lived a life in which their values were the same: "Living well is the best revenge" was their favorite saying, and their lifestyle celebrated that injunction. They were happy in their lifestyle of enjoying the moment, but now an issue of basic importance was threatening to destroy that lifestyle. They were feeling they were living in terror-time: their old consensus of how to live their marriage was being destroyed. And if they remained in total disagreement about whether or not to have a child, they were headed for an abyss. There is great comfort in the familiar, and when a familiar lifestyle is threatened with destruction, fear of an unknown future can become an abyss of terror.

That was the basic reason for the wild and bitter accusations they directed toward each other. After all, they were two sophisticated adults, so why couldn't they solve this problem that divided them in a rational, nonjudgmental way? They couldn't because the issue as to whether or not to have a child was simply the cover story for the terror they were feeling that was hidden beneath this disagreement.

Age, the process of becoming older, was the fear that was dividing them. The Silicon Valley virus that infected the information-technology community—the virus of believing being young is the only time in life that truly mattered; the time in life that the poet William Wordsworth exalted as "very heaven"—had infected Carol and Harry. Silicon Valley was awash with hundreds of thousands of information technologists who had just graduated from colleges, business schools, and universities in the 1990s. Their technical knowledge was in high demand at that time and their age (their mid-twenties usually) was regarded by employers as an outstanding advantage when it came to hiring them. For these were the men and women who had the energy and the drive to succeed at all costs; they also were unattached and even if married had little or no outside commitments to interfere with jobs that demanded eighteen or more hours a day, and they were always on call outside of working hours. They had no family life to interfere with the projects they were working on since their families of origin usually lived in states other than California or in a foreign country (almost half of skilled Silicon Valley employees came from India). Youth was the gold mine CEOs could exploit, and the media hype of the 1990s kept proclaiming that not only was Silicon Valley a revolutionary new industry, it was an industry whose revolution was powered by young people. It was a time when everyone over thirty-five or forty was considered expendable. The corporations of Silicon

Valley regarded workers in their late thirties and forties as "over the hill," expendable drains on profit. Loyalty and trust and the value of experience may have been appropriate for the old manufacturing way of life, but this new century was the Information Age century in which hiring and firing in obeisance to the bottom line took precedence over the other values in life.

As long as one was in one's twenties, this devaluation of becoming older was accepted as something to brag about and flaunt in the face of people in their thirties or forties who were doomed to live on a fraction of what the young men and women in the Valley were making. They had bought the belief that after thirty, creativity is lost. After all, that was Bill Gates' statement in that era (before he grew older!). They were entitled to their economic success *because* they were young was the thinking of that decade.

Carol and Harry had bought this illusion that being young was the only value in life that mattered. But this illusion began to erode when they turned thirty. They made excuses for themselves on their thirtieth birthdays that they were not really deteriorating, going down hill in life after all. They seemed, however, to be protesting too much, that they were whistling-in-the-dark and really believed life would never be as good for them as it had been in their twenties. Harry fought this feeling in every conceivable way. He still ran five miles each morning, swam in his pool as often as he did in his twenties, worked out at the gym, and prided himself as being no heavier than he was when he first came to Silicon Valley in the early '90s. He had inherited a healthy head of hair and was relieved to know he would never become bald. He hadn't even a single gray hair on his head! He looked as if he was still in his late twenties when Pat and I met him. He smiled when he told us he was thirty-five going on thirty-six, as if to say, "Can you believe it? You could never mistake me for a guy in his mid-thirties!" Here it was the middle of 2001, and he was still living in the age of fantasy of the mid-1990s.

Carol began to reluctantly let go of that fantasy of being young forever when she saw the first wrinkles around her eyes and strands of gray hair that she had to dye because there were too many to pluck out. She was wrestling with the signs that she had left her twenties forever a year before she decided to have a child. It was also at that time that she worried about the threat of menopause. Both her mother and grandmother had their menopause early in life—thirty-nine for her mother, thirty-eight for her grandmother. So the odds were likely that she, too, would have an early menopause, even before she turned forty. It was so unfair. She had felt as if she were the only person in the world singled out to become older. When she told us that was how she felt at that time, she laughed sheepishly, acknowledging it was such an absurd attitude. But that was how she felt about growing older. While these feelings about her growing older were developing, she began to surprise herself by toying with the idea of having a child. It had started when she was thirty-two. She found she was surprising herself by noticing babies in their baby carriages. She paid attention to the women who were holding the hands of their young children as they shopped in the supermarket. She was suffused with warm feelings as she saw mothers or fathers delight in playing with their four- or five-year-olds in a neighborhood playground.

These emerging feelings created a jumble of contradictory thoughts in her mind. On the one hand, she felt guilty about going back on her word never to have children, for here she was thinking that it might be nice to have a child. She remembered how outspoken she had been about wanting to be free of any obligations that would tie her down to a life of dirty diapers and screaming kids who would eat up all of her free time. No way!

But that was then, in her twenties, when she was having unfettered fun and experiencing the world in long travel trips. She had been there and done that; doing more of the same the rest of her life now held little attraction for her. She seemed to be ready for something else. And that something else was thinking about having a child. She had started to change her ideas about children. Sure, they could be a pain in the neck, but they could be gratifying also. The thought of creating something of her own—to have it grow from a seed to a live human being, and then give birth to that baby and respond to the challenge of nurturing its growth—how wonderful that would be!

This was the new turn in her thinking that she never had shared with Harry. All Harry knew was her abrupt statement that she wanted to have a child. Period. Out of the blue. In the counseling sessions she explained to Harry that she was frightened of explaining her feelings to him because she felt that he might not understand them and take them as an excuse to accuse her of not being true to her marriage commitment never to have a child. And she now saw she was right.

The fact was that, yes, she had gone back on her word. But now she did not see it as a betrayal of Harry. There was no need for her to feel guilty, for she had come to terms with the fact that people change and that the person who committed herself to one point of view in her twenties was *not* the same person who changed that point of view in her thirties. There was no betrayal involved, since Carol at thirty-five was *not* Carol at twenty-two.

Carol and Harry had shared a good life together which was why they came to see us. They really wanted to save their marriage because there was a firm grounding of caring for each other that still resided in their relationship. But in their three months of once-a-week counseling with us, marriage counseling had turned into divorce counseling, for they now realized that what had brought them together in the first place—common interests, common values, and sexual chemistry—no longer existed. Carol had changed, but Harry remained the same twenty-year-old she had married. Harry's mind was closed. He had no sympathy for the idea that Carol was a changed person and therefore had a right to change her mind. He kept repeating in the counseling sessions that once a commitment as basic as the commitment not to have a child was broken, it was pure and simple a betrayal of all their marriage stood for. He would consider the possibility of trying to renew their marriage only if Carol gave up this "stupid" idea of wanting a child.

Carol replied that he was being unfair and inconsiderate and had no respect for her feelings. No, she would never agree to stay in the marriage if they didn't have a child.

Having a child meant acknowledging you were growing older. But Harry wanted to be the only child in his marriage to

Carol. For his behavior throughout the marriage was the behavior of a self-indulgent child, always wanting center stage, always wanting instant gratification. He considered his impulsive behavior a badge of honor, not something to be modified. These character traits, combined with his very competent adult business abilities, made him attractive to many people. He created an image of boyish charm combined with adult business success.

Carol stated she now realized that Harry would always remain an adult boy, but what she wanted was to share a life with a partner who also would become a parent.

Carol cried and there were tears in Harry's eyes when they both decided the time had come to turn their counseling sessions with us into divorce mediation. They made a very admirable effort to be kind and fair to each other. Neither of them used the divorce negotiations over property as an excuse to exact vengeance from the other partner for the emotional hurt each was experiencing. They had invested almost a third of their lives in their marriage and chose to honor that investment by wishing each other well as they moved in separate directions.

Carol decided to leave Silicon Valley. It had been a source of economic well-being and excitement in her twenties, but it had worn out its welcome in her heart and mind now that she was in her mid-thirties. She was now open to other change in her life, such as changing her career as well as the community she lived in. Boston appealed to her as a grown-up town she had enjoyed visiting previously. She decided to live there and change her career to become a hospital administrator. She wanted to remarry, but it would be because she loved the person, not just because she wanted a man to father a child. In a recent letter to us, she said she had not found that person, but was prepared to have a child through artificial insemination if time was running out and love took longer to find than she hoped it would.

Harry remained Harry. He still lives in Silicon Valley, still is successful in the job he still likes. He's known as a "player" and enjoys surreal short-term relationships. He continues to fight the battle of staying young forever and is now busy looking into new genetic and hormonal developments as possibilities for arresting the aging process. He has yet to realize that only the dead stay young forever.

5. THE DISPOSSESSED

The novelist Alexander Kuprin once remarked, "The horror is that there is no horror!" It's a profound statement. In eight words it sums up the way human beings have a seemingly natural tendency to overlook or ignore the horrible things that are right before their very eyes. In Silicon Valley, the area is rife with thousands of homeless, unemployed or working poor men and women. But all eyes are focused instead on the middle- and upper-class lifestyles that glisten in each high-tech city. One of the wealthiest areas in the wealthiest country in the world contains a permanent underclass that is regarded as nonexistent. Yet this archetype of Silicon Valley is as permanent as part of its culture as the vaunted high-tech men and women who parade through its streets.

This underclass has best been described in the following way:

> The poor inhabit a world scarcely recognizable, and rarely recognized, by a majority of their fellow Americans. It is a world apart, whose inhabitants are isolated from the mainstream of American life and alienated from its values. It is a world where Americans are literally concerned with day-to-day survival, a roof over their heads, where the next meal is coming from. It is a world where a minor illness is a major tragedy, where pride and privacy must be sacrificed to get help, where honesty can become a luxury and ambition a myth!

The more things change, the more they remain the same. For this snapshot of the dispossessed was written in a national government report by the Council of Economic Advisers in January 1964 (p. 55)!

Must the dispossessed remain a permanent feature of the American landscape? The movers and shakers of Silicon Valley have repeatedly proclaimed that they have the answer to all of society's major problems. They need only to look a small distance beyond their own backyard to isolate the problems of their own dispossessed and solve them.

SEPTEMBER 11,
2001
"ALL THAT IS SOLID MELTS INTO AIR"

I t's as if Shakespeare were alive and witnessed New York's twin towers of the World Trade Center disappear in two hours and came up with the right words to describe what happened. The solid architecture designed to last for many lifetimes became a hole in the sky.

And as if this traumatic event were not sufficient to turn our world upside down, the active war began on October 7, 2001, and released savage and terrifying new uncertainties.

The law of unintended consequences is operating overtime: our high-tech, "invulnerable" society became vulnerable with low-tech explosions from two commercial airlines. Our superprotected Pentagon turned out to be unprotected, after all. Suddenly, the very ground we stand on feels like quicksand: the water supply we live by, the air we breathe, and the elimination of epidemics long taken for granted, all are now at risk with biological and chemical warfare and ordinary bombs that can change life into death at a moment in time.

No one can predict our future. The "icons" of Silicon Valley have been silent. What can they say? They always had tunnel vision in the first place, obsessed as they were with the bottom lines and stock valuations, with being number one in the marketplace the only thing that mattered.

These icons of industry always talked of their competition as being enemies in a battle for survival. They talked about the "art of war" repeatedly as if that art guaranteed their success in life. Now they have (like all of us) a *real* war in which survival depends on other, infinitely more important, factors than the fine print in the *Wall Street Journal*.

In what seems like the blink of an eye, our priorities for our country and ourselves have taken a 180-degree turn since September 11, 2001. The old Silicon Valley belief, accepted by the general public, that information technology would change the world into a garden of peace and prosperity forever has vanished. Before we lay that delusion to rest, it is well to recall a typical statement made by a high-tech enthusiast before about the virtues of high technology:

"Technology and the Net, used creatively, can bring people together in ways that have never been done before. Artificial barriers and controls are in the brink of extinction, thanks to innovative and intelligent applications of technology. With a populace that is informed, enthusiastic and open to new ideas, old-style oppression will be exposed almost as soon as it's applied" (AlterNet, July 31, 2001).

Similar statements were voiced like mantras throughout Silicon Valley during the years when the stock market seemed to be creating the utopia the information technologists wished for. This naive belief in the superior virtue of technological innovation as the key to a secure and happy world is nothing new. For example, here is another example from another time in history:

"This was to have been remembered as the century of perfected human communications—of swift air mail letters flying over oceans and lands, of radio stations comfortably crackling sparks of news into the night, of wireless-telephoned headlines presumably announcing that mankind was all well, and of streamlined trains hustling across continents with unimportant postcards. . . ."

These words were written on December 7, 1940, by *New Yorker* reporter Janet Flanner, then in occupied France. World War II was raging at that time. Philosopher George Santayana's warning that "those who don't remember the past are condemned to repeat it" is the appropriate observation to make about this search for a nonexistent technological utopia.

It is an overlooked fact that with the advent of the new technology, beginning with the invention of the transistor shortly after World War II, and with the accelerated advance of the technologies that was supposed to better the world in the subsequent decades, war has been technology's partner, not peace. James Meek in the *London Review of Books*, September 6, 2001, itemizes this roll call of death:

It's not as if there haven't been any wars since 1945. There have been about three hundred. It's true that people haven't fought and died everywhere in the world. They've fought and died only in Indonesia, Greece, Iran, Vietnam, India, Bolivia, Pakistan, China, Paraguay, Yemen, Madagascar, Israel,

Colombia, Costa Rica, Korea, Egypt, Jordan, Lebanon, Syria, Burma, Malaysia, Colombia, the Philippines, Thailand, Tunisia, Kenya, Taiwan, Morocco, Guatemala, Algeria, Argentina, Cameroon, Hungary, Haiti, Rwanda, Sudan, Oman, Honduras, Nicaragua, Mauritania, Cuba, Venezuela, Iraq, Zaire, Laos, Burundi, Guinea-Bissau, Somalia, France, Cyprus, Zambia, Gabon, the US, Uganda, Tanzania, Brazil, the Dominican Republic, Peru, Namibia, Chad, Czechoslovakia, Spain, the Soviet Union, Britain, El Salvador, Cambodia, Italy, Sri Lanka, Bangladesh, Chile, Turkey, Ethiopia, Portugal, Mozambique, South Africa, Libya, Afghanistan, Jamaica, Ghana, Ecuador, Zimbabwe, Burkina Faso, Mali, Panama, Romania, Senegal, Kuwait, Armenia, Azerbaijan, Niger, Croatia, Georgia, Bhutan, Djibouti, Moldova, Sierra Leone, Bosnia, Tajikistan, the Congo, Russia, Mexico, Nepal, Albania, Yugoslavia, Eritrea, Macedonia and Palestine. One crude tally puts the number of dead at well over twenty million. . . .

The title of our book is *Down and Out in Silicon Valley*. Since our book deals with the quality of people's lives, "Down and Out" does not mean economic deprivation: It signifies the barrenness, the "Is-This-All-There-Is?" feelings that have dominated the existence of the tens of thousand of men and women who came to Silicon Valley "for the money." This single-minded pursuit became a dead end before, as well as after, the recession hit the Valley. Our subtitle, "The High Cost of the High-Tech Dream" is the personal price paid for sacrificing one's entire life to the goal of becoming wealthy, only to discover that he or she has lost his or her sense of self in the process of "making it."

For a moment in time (1995–2000), the Silicon Valley dream became a national obsession: Millions of hardworking people working in factories, offices, and the service industries believed what the media was telling them (for example, "You're just an ordinary worker, but you, too, can become rich by playing the stock market. Invest in high-tech stocks and you will be wealthy overnight!"). So these hardworking people invested money they could not afford in the hope of making a quick fortune. They would become day traders, with eyes glued to a computer screen for endless hours each day; they would wake up in the morning, and without paying attention to their spouse or children, instantly click on CNBC's *Market Watch*, and anxiously check out their stock prices. They were unaware that making money in the stock market is an insider's game—and that they were only victimized outsiders. Pat and I counseled many of these men and women who were caught up in this frenzy of speculation, which caused anxiety and depression and left families in bitter disarray. More psychic and economic devastation occurred in these investors than in the young Silicon Valley men and women who lost their jobs and income when the Silicon Valley bubble burst. For most of those high-tech engineers, merchandisers, marketers, and human resource workers came from comfortable middle- and upper-middle-class households and could return to a nurturing family and start over in any new direction they chose. The average investor, on the other hand, had no such lifeline available.

The recession that began in mid-2000 became the beginning of much soul-searching in our society. Since the quality of life based solely on the search for more and more wealth was tested and found wanting, what was the alternative? Even the headhunting Web sites were responding to this new trend: Advertisements were appearing that proclaimed that 75,000 jobs were available—jobs where you did not have to lose your soul if you accepted one of them. Not "losing your soul" became a more important selling point than salary.

Paradoxically, the stage was being set for a reassessment of the values one needed to live an enriching life. Yes, money was a factor, but only *one* factor in a life. In fact, there were visionaries who cautioned CEOs and other high-tech elites as early as 1998 that "IPOs are well and good. But the real question is: How are your legacies? How are you loving all your children?" This was said at a TED Conference (Technology, Entertainment, and Design) before an audience of high-tech CEOs and their elite subordinates by Professor William H. McDonough of the University of Virginia. He spoke this gentle warning on July 27, 1998, but the audience then was more interested in the fabulous stock market at that time than in a loving family relationship.

The most dramatic indication that high-tech success is not the key to happy family life came from an astounding speech by Bill Gates, the Microsoft giant, who stated at a "Creating Digital Dividends" conference in November 2000, that he had been "naive—very naive" for believing that the computer industry could solve the major problems facing humanity, since the billions of people who subsist on a dollar a day are not in a position to benefit from the Information Age. The *New York Times*, reporting his speech, stated that Gates "has lost much of the faith he once had that global capitalism would prove capable of solving the most immediate catastrophes facing the world's poorest people, especially the 40,000 deaths a day from preventable diseases. He added that more philanthropy and more government aid—especially a greater contribution to foreign health programs by American taxpayers—are needed for that."

Today, Silicon Valley is a different place. No, it is not a sequence of ghost towns, but it is a less ebullient, more somber, less arrogant, more reflective place to work and live in. September 11, 2001, had a shattering effect on its economy: More layoffs, more red-ink bottom lines, and great uncertainty about sales, since people are no longer as entranced with new computer developments as they once were.

The Ayn Rand philosophy of selfishness no longer holds the appeal it once had. More important values are at stake, such as survival as a country, survival as an economy, survival as an individual. The terrorist attack on the twin New York towers and on the Pentagon have become a testing arena for self-examination. The Silicon Valley people Pat and I counsel often mention two sentences: The first is that after September 11, they believe "There are things that are much, much more important than the bottom line." The second is: "If you want to make God laugh, make plans for tomorrow."

Silicon Valley arrogance is being superseded by a genuine concern for others, as well as for oneself. The *Mercury News*, the outstanding newspaper that is the voice of Silicon Valley, is echoing this new fact of life. It featured on September 27, 2001, an article that acknowledged this new sense that cooperation and kindness and compassion are the survival values that need to be affirmed. The article, written by Sue Hutchison, stated "Becoming better citizens may be the best way to honor September 11 dead. . . . This tragedy may be an opportunity for Americans to reorganize our priorities and become citizens of the world in a way many of us have never been before. That may be the best way to honor those who died. . . ."

Silicon Valley is very much alive. It has withstood many transformations in the past, as our previous chapters noted. But it always has had the resilience to turn a corner and come out triumphantly in the next turn of the economy. While it would be foolish to predict the specific ways in which Silicon Valley will grow again, there is a general indication of the direction in which it will develop. David Lazarus, writing in the *San Francisco Chronicle* on September 23, 2001, had some perceptive comments about the direction the renewal of Silicon Valley is taking:

> Silicon Valley insiders and Wall Street analysts say a torrent of new spending is about to be unleashed as demand surges for technologies capable not just of protecting national resources but also serving as the spear tip in America's efforts to strike back against shadowy foes.
> The battlefield will not be physical so much as it will be

digital," said Rob Owens, a technology industry analyst at Pacific Crest Securities in Portland, Ore. 'There will definitely be people who prosper in this new environment.'

Among developments to watch:

▼ A stronger focus on research and development as Silicon Valley companies target more secure technologies for Internet traffic and computer networks, as well as core components, such as microchips, for sophisticated weaponry.

▼ An increase in defense spending among biotech firms as the nation girds for potentially more lethal attacks involving chemical and biological weapons.

▼ A new lease on life for hard hit telecom companies as a wariness of flying spurs a renewed interest in video conferencing and other means of electronic gathering.

▼ A gradual increase in sales across the tech spectrum as customers upgrade hardware and software systems to accommodate the needs of a more hostile environment.

"There will be a rising-tide effect in the tech world," said Rob Batchelder, research director at the Gartner Group in Stamford, Conn. "There's no question about it."

The father of all modern science, without whose discoveries Silicon Valley would be nonexistent, should have the last word about how to survive as human beings in light of the life-and-death struggle our country is experiencing. Albert Einstein said:

> "The ideals which have lighted my way, and time after time have given me new courage to face life cheerfully, have been Kindness, Beauty and Truth. . . . The trite subjects of human efforts—possessions, outward success, luxury—have always seemed to me contemptible."